Course **1**

GLENCOE **MATH**

*YOUR COMMON CORE EDITION* → CCSS

 **Assessment Masters**

WITHDRAWN

 **Education**

*Bothell, WA • Chicago, IL • Columbus, OH • New York, NY*

connectED.mcgraw-hill.com

Education

**STEM** McGraw-Hill is committed to providing
instructional materials in Science, Technology,
Engineering, and Mathematics (STEM) that give all
students a solid foundation, one that prepares them for
college and careers in the 21st century.

Send all inquiries to:
McGraw-Hill Education
8787 Orion Place
Columbus, OH 43240

ISBN: 978-0-07-662327-3
MHID: 0-07-662327-0

Printed in the United States of America.

9 10 11 12 13 QVS 19 18 17 16 15

 *Assessment Masters*

Our mission is to provide educational resources that enable
students to become the problem solvers of the 21st century
and inspire them to explore careers within Science, Technology,
Engineering, and Mathematics (STEM) related fields.

# Teacher's Guide to Using the
## *Assessment Masters*

The *Assessment Masters* includes the core assessment materials needed for each chapter. The answers for these pages appear at the back of this book.

## Are You Ready? Worksheets

- Use after the Are You Ready? section in the Student Edition.

## Chapter Diagnostic Test

- Use to test skills needed for success in the upcoming chapter.
- Retest approaching-level students after the Are You Ready? worksheets.

## Chapter Pretest

- Quick check the upcoming chapter's concepts to determine pacing.
- Use before the chapter to gauge students' skill level and to determine class grouping.

## Chapter Quiz

- Reassess the concepts tested in the Mid-Chapter Check in the Student Edition.

## Vocabulary Test

- Includes a list of vocabulary words and questions to assess students' knowledge of those words and can be used in conjunction with one of the Chapter Tests.

## Standardized Test Practice

- Assess knowledge as student progresses through the textbook.
- Includes multiple-choice, short-response, gridded-response, and extended-response questions
- Student Recording Sheet corresponds with the Test Practice.

## Extended-Response Test

- Contains performance-assessment tasks and includes a scoring rubric

## Chapter Tests

- **AL** 1A-1B Approaching-level students; contains multiple-choice questions
- **OL** 2A-2B On-level students; contains both multiple-choice and free-response questions
- **BL** 3A-3B Beyond-level students; contains free-response questions
- Tests A and B are created with parallel format. Use when students are absent or for different rows.

## Benchmark Tests

- Contains multiple-choice and short-response questions
- The first three tests provide quarterly evaluations.
- The last test provides a cumulative end-of-year evaluation.

# CONTENTS

## Chapter 5 Integers and the Coordinate Plane

## Chapter 6 Expressions

## Chapter 7 Equations

## Chapter 8 Functions and Inequalities

# Chapter 9 Area

# Chapter 10 Volume and Surface Area

# Chapter 11 Statistical Measures

# Chapter 12 Statistical Displays

# Are You Ready?

## Review

### Example 1

Find $6\overline{)486}$.

$$
\begin{array}{r}
81 \\
6\overline{)486} \\
-48 \\
\hline
6 \\
-6 \\
\hline
0
\end{array}
$$

Divide each place-value position from left to right.

Since $6 - 6 = 0$, there is no remainder.

### Example 2

Find $12\overline{)276}$.

$$
\begin{array}{r}
23 \\
12\overline{)276} \\
-24 \\
\hline
36 \\
-36 \\
\hline
0
\end{array}
$$

Divide each place-value position from left to right.

Since $36 - 36 = 0$, there is no remainder.

## Exercises

### Divide.

1. $4\overline{)80}$

2. $6\overline{)72}$

3. $5\overline{)430}$

4. $8\overline{)224}$

5. $15\overline{)390}$

6. $14\overline{)252}$

7. $41\overline{)492}$

8. $37\overline{)629}$

1. _____

2. _____

3. _____

4. _____

5. _____

6. _____

7. _____

8. _____

# Are You Ready?

## Practice

**Divide.**

1. $4\overline{)76}$

2. $7\overline{)91}$

3. $15\overline{)165}$

4. $61\overline{)366}$

5. **AIR SHOW** Joel bought 6 tickets to an air show. If he spent $156, how much did each ticket cost?

6. **MEASUREMENT** To visit family, Mr. Yusef drove 297 miles in 3 days. How many miles did he travel each day on average?

**Write each fraction in simplest form.**

7. $\dfrac{17}{51}$

8. $\dfrac{14}{18}$

9. $\dfrac{20}{90}$

10. $\dfrac{48}{56}$

11. **TREES** Of the 84 trees in a nursery, 33 are orange trees. What fraction, in simplest form, of the trees are orange trees?

12. **MARBLES** Julio bought 12 new marbles, bringing his total number of marbles to 72. What fraction, in simplest form, of Julio's marbles are new?

1. _____

2. _____

3. _____

4. _____

5. _____

6. _____

7. _____

8. _____

9. _____

10. _____

11. _____

12. _____

# Are You Ready?

## Apply

| | |
|---|---|
| **1. FIELD TRIP** A group of 92 students went on a field trip to a nature preserve. When they arrived, they were separated into 4 groups. If each group had the same number of students, how many students were in each group? | **2. CARS** The 91 new cars in a dealership are arranged into 7 rows. If each row has the same number of cars, how many cars are in each row? |
| **3. EARNINGS** Ming earned $336 from babysitting over the past 6 weeks. She earned the same amount of money each week. How much money did she earn each week? | **4. GASOLINE** The Harnett family used 18 of the 26 gallons of gasoline they purchased. What fraction of the gasoline, in simplest form, did they use? |
| **5. CAR WASH** The student council waxed 27 of the 63 cars they washed. What fraction of the cars, in simplest form, did they wax? | **6. BARBEQUE** Mr. Salcido bought 24 hot dogs and 36 hamburgers for a barbeque. What fraction, in simplest form, of his food items are hot dogs? |

# Diagnostic Test

**Divide.**

1. $2\overline{)58}$

2. $4\overline{)76}$

3. $42\overline{)882}$

4. $31\overline{)961}$

5. **CLOTHES** Mrs. Nicci had $228 to spend on school clothes for her 3 children. If she spent the same amount of money on each child, how much money did she spend on each child?

**Write each fraction in simplest form.**

6. $\dfrac{14}{70}$

7. $\dfrac{5}{95}$

8. $\dfrac{63}{81}$

9. $\dfrac{12}{60}$

10. **CEREAL** Of the 35 types of cereal sold in a store, 15 are advertised to be whole grain. What fraction, in simplest form, of the cereals are advertised as whole grain?

1. _____

2. _____

3. _____

4. _____

5. _____

6. _____

7. _____

8. _____

9. _____

10. _____

# Pretest

**Write each ratio as a fraction in simplest form.**

1.

   baseballs to basketballs

2. 12 kittens to 15 puppies

3. **CLOTHING** At the ticket booth at a Florida Marlins game, 15 people wore turquoise and 24 people wore gray. What is the ratio of people who wore turquoise to the people who wore gray in simplest form?

4. **MUSIC** Mangi surveyed students about their favorite types of music. Find the ratio of students who chose dance to the total number of students surveyed.

   | Favorite Music | |
   |---|---|
   | Pop | 7 |
   | Country | 5 |
   | Dance | 4 |

5. Write 90 words in 2 minutes as a unit rate.

6. **GRASS SEED** Mr. Ernesto spent $72 for 3 bags of grass seed. How much did he spend on each bag?

**Use the ratio table to solve the problem.**

7. **MIXTURE** A mixture requires 4 cups of water for every 7 cups of flour used. How many cups of water should be used with 28 cups of flour?

   | Cups of Flour | 7 | 28 |
   |---|---|---|
   | Cups of Water | 4 | ■ |

1. _____

2. _____

3. _____

4. _____

5. _____

6. _____

7. _____

# Chapter Quiz

1. Find the greatest common factor of 13 and 39.

2. Find the least common multiple of 3 and 21.

**For Exercises 3 and 4, write each ratio as a fraction in simplest form. Then explain its meaning.**

3.

    cats to dogs

4. There are 16 girls and 20 boys in the school band. What is the ratio of girls to boys?

**Write each rate as a unit rate.**

5. $8 for 2 pounds of rice

6. 240 miles in 6 hours

7. Chin bought 12 pounds of apples for $6. Use the ratio table to find the cost of buying 3 pounds of apples.

| Pounds of Apples | 12 | | 3 |
|---|---|---|---|
| Price of Apples | $6 | | ■ |

8. **MONEY** The table shows the amount of money Dawson earned for watching his neighbor's cat. Graph the ordered pairs (number of days, amount earned). Then describe the pattern in the graph.

| Amount Earned ($) | 0 | 7 | 14 | 21 |
|---|---|---|---|---|
| Number Of Days | 0 | 1 | 2 | 3 |

9. Markus receives dollar bills from his grandfather for one week. On the first day he received $1; on the second day he received $2, on third day he received $4, and on the fourth day he received $8. If this continues, how much money will Markus have by the end of the week?

1. _____

2. _____

3. _____

4. _____

5. _____

6. _____

7. _____

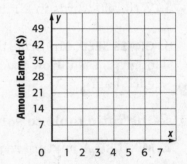

8. _____

9. _____

**6**

**Course 1 • Chapter 1** Ratios and Rates

# Vocabulary Test

| | | | |
|---|---|---|---|
| coordinate plane | least common multiple | rate | unit rate |
| equivalent ratio | ordered pair | ratio | x-axis |
| graph | origin | ratio table | x-coordinate |
| greatest common factor | prime factorization | scaling | y-axis |
| | | unit price | y-coordinate |

**Choose from the terms above to complete each sentence.**

1. _____ express the same relationship between two quantities.

1. _____

2. The rate for one unit of a given quantity is called a(n) _____.

2. _____

3. Multiplying or dividing two related quantities by the same number is called _____.

3. _____

4. In a _____, the columns are filled with pairs of numbers that have the same ratio.

4. _____

5. The least number that is a multiple of two or more whole numbers is _____.

5. _____

**Choose the correct term.**

6. In an ordered pair, the (x-coordinate, y-coordinate) is listed first.

6. _____

7. The (least common multiple, greatest common factor) of 5 and 10 is 5.

7. _____

8. In the ordered pair (2, 3), 3 is the (x-coordinate, y-coordinate).

8. _____

**Define each term in your own words.**

9. ratio

9. _____

10. rate

10. _____

# Standardized Test Practice

**Read each question. Then fill in the correct answer on the answer sheet provided by your teacher or on a sheet of paper.**

1. The ratio of cats to dogs seen by a veterinarian in one day is 2 to 5. If a vet saw 40 dogs in one day, how many cats did he see?

   A. 5             C. 29
   B. 16            D. 40

2. At a sports camp, there is one counselor for every 12 campers. If there are 156 campers attending the camp, which equivalent ratios could be used to find the number of counselors?

   | Campers | Counselors |
   |---------|-----------|
   | 12      | 1         |
   | 24      | 2         |
   | 156     | ■         |

   F. $\dfrac{\blacksquare}{12} = \dfrac{1}{156}$     H. $\dfrac{12}{1} = \dfrac{\blacksquare}{156}$

   G. $\dfrac{1}{12} = \dfrac{\blacksquare}{156}$     I. $\dfrac{\blacksquare}{1} = \dfrac{12}{156}$

3. ✎ **GRIDDED RESPONSE** A car jack requires a force of 110 pounds to lift a 2,500-pound car. In simplest form, what is the ratio of the car's weight to the force required to lift the car?

4. A baker is making an oversized sheet cake for a school dance. The recipe calls for 2 cups of sugar for every 3 cups of flour. How many cups of flour are needed, if she is using 18 cups of sugar?

   A. 27 cups
   B. 18 cups
   C. 9 cups
   D. 6 cups

5. Which rate gives the best price for scrapbook paper?

   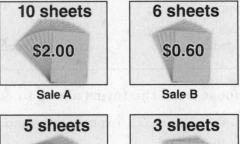

   | 10 sheets | 6 sheets |
   |-----------|----------|
   | $2.00     | $0.60    |
   | Sale A    | Sale B   |

   | 5 sheets  | 3 sheets |
   |-----------|----------|
   | $1.00     | $0.60    |
   | Sale C    | Sale D   |

   F. Sale A        H. Sale C
   G. Sale B        I. Sale D

6. Which of the following ratios is equivalent to $\dfrac{5}{7}$?

   A. $\dfrac{10}{12}$     C. $\dfrac{15}{17}$

   B. $\dfrac{3}{5}$       D. $\dfrac{10}{14}$

7. ✎ **GRIDDED RESPONSE** The table shows the cups of whole wheat flour required to make dog biscuits.

   | Number of Dog Biscuits | 10 | 35 |
   |------------------------|----|----|
   | Cups of Whole Wheat Flour | 2 | ■ |

   How many cups of whole wheat flour are required to make 35 biscuits?

8. A school population was predicted to increase by 50 students a year for the next 10 years. If the current population is 700 students, what will be the enrollment after 10 years?

   F. 50 students       H. 1,200 students
   G. 500 students      I. 7,000 students

9. **SHORT RESPONSE** There are 12 boys and 15 girls in your class. Four new students, 3 boys and 1 girl, join your class. Compare the ratio of boys to girls before and after the new students enrolled.

10. The table shows the cost of different kinds of fruit.

| Fruit | banana | apple | orange | grapes |
|---|---|---|---|---|
| Cost per pound ($) | 0.60 | 1.89 | 0.99 | 1.00 |

To the nearest cent, what is the cost of 3.5 pounds of grapes?

**A.** $2.84    **C.** $3.50

**B.** $2.88    **D.** $3.63

11. **GRIDDED RESPONSE** Mr. Reinard bought 5 pounds of feed for his chickens for $10. How much did the feed cost per pound in dollars?

12. **SHORT RESPONSE** The length of a track around a football field is $\frac{1}{4}$ mile How many miles do you walk if you walk $2\frac{3}{4}$ times around the track?

13. The bar graph below shows the average number of visitors to different Web sites during one month.

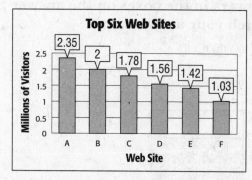

How many visitors can Web site B expect to have in one year?

**F.** 2.4 million

**G.** 24 million

**H.** 240 million

**I.** 2,400 million

14. **EXTENDED RESPONSE** Cesar's sixth-grade class sorted books in the library. The class sorted 45 books in 90 minutes.

*Part A* Write an equivalent ratio to find how long it would take to sort 120 books.

*Part B* How many hours will it take the class to sort 120 books?

*Part C* Suppose their rate slowed to 30 books in 90 minutes. How long would it take the class to sort the 120 books? Explain your reasoning.

# Student Recording Sheet

Use this recording sheet with the Standardized Test Practice.

Fill in the correct answer. For gridded-response questions, write your answers in the boxes on the answer grid and fill in the bubbles to match your answers.

1. Ⓐ Ⓑ Ⓒ Ⓓ

2. Ⓕ Ⓖ Ⓗ Ⓘ

3.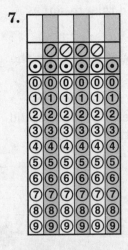

4. Ⓐ Ⓑ Ⓒ Ⓓ

5. Ⓕ Ⓖ Ⓗ Ⓘ

6. Ⓐ Ⓑ Ⓒ Ⓓ

7.

8. Ⓕ Ⓖ Ⓗ Ⓘ

9. _____

10. Ⓐ Ⓑ Ⓒ Ⓓ

11.

12. _____

13. Ⓕ Ⓖ Ⓗ Ⓘ

## Extended Response

Record your answers for Exercise 14 on the back of this paper.

# Extended-Response Test

Demonstrate your knowledge by giving a clear, concise solution to each problem. Be sure to include all relevant drawings and justify your answers. You may show your solution in more than one way or investigate beyond the requirements of the problem. If necessary, record your answer on another piece of paper.

1. **a.** Tell in your own words the meaning of *ratio*.

   **b.** Give an example of a ratio. Write the ratio in four ways.

   **c.** Tell in your own words the meanings of *rate* and *unit rate*. Give an example of a unit rate and an example of a rate that is not a unit rate.

   **d.** Tell in your own words the meaning of *equivalent ratios*.

   **e.** Write a word problem that you can solve by using equivalent ratios.

   **f.** Solve the word problem in part **e**. Explain each step.

2. **FINANCIAL LITERACY** The U.S. dollar can be exchanged for other types of currency in the world. The exchange rate is always changing. Suppose 1 euro is equivalent to $2.

| Euro | 1 | 2 | 5 | 10 |
|---|---|---|---|---|
| U.S. Dollar ($) | 2 | | | |

   **a.** Describe what is meant by the numbers in the first column of the table.

   **b.** How much will 2 euros cost in U.S. dollars? 5 euros? 10 euros?

   **c.** How many euros is equivalent to $30? Tell how you found your answer.

   **d.** List the information in the table as ordered pairs (euro, U.S. dollar).

# Extended-Response Rubric

| Score | Description |
|:---:|---|
| 4 | A score of four is a response in which the student demonstrates a thorough understanding of the mathematics concepts and/or procedures embodied in the task. The student has responded correctly to the task, used mathematically sound procedures, and provided clear and complete explanations and interpretations.<br><br>The response may contain minor flaws that do not detract from the demonstration of a thorough understanding. |
| 3 | A score of three is a response in which the student demonstrates an understanding of the mathematics concepts and/or procedures embodied in the task. The student's response to the task is essentially correct with the mathematical procedures used and the explanations and interpretations provided demonstrating an essential but less than thorough understanding.<br><br>The response may contain minor flaws that reflect inattentive execution of mathematical procedures or indications of some misunderstanding of the underlying mathematics concepts and/or procedures. |
| 2 | A score of two indicates that the student has demonstrated only a partial understanding of the mathematics concepts and/or procedures embodied in the task. Although the student may have used the correct approach to obtaining a solution or may have provided a correct solution, the student's work lacks an essential understanding of the underlying mathematical concepts.<br><br>The response contains errors related to misunderstanding important aspects of the task, misuse of mathematical procedures, or faulty interpretations of results. |
| 1 | A score of one indicates that the student has demonstrated a very limited understanding of the mathematics concepts and/or procedures embodied in the task. The student's response is incomplete and exhibits many flaws. Although the student's response has addressed some of the conditions of the task, the student reached an inadequate conclusion and/or provided reasoning that was faulty or incomplete.<br><br>The response exhibits many flaws or may be incomplete. |
| 0 | A score of zero indicates that the student has provided no response at all, or a completely incorrect or uninterpretable response, or demonstrated insufficient understanding of the mathematics concepts and/or procedures embodied in the task. For example, a student may provide some work that is mathematically correct, but the work does not demonstrate even a rudimentary understanding of the primary focus of the task. |

# Test, Form 1A

**Write the letter for the correct answer in the blank at the right of each question.**

1. What is the least common multiple of 6 and 8?

   **A.** 12      **B.** 24      **C.** 32      **D.** 48      1. _____

2. At a summer camp, there are 40 boys out of 70 campers. What is this ratio written as a fraction in simplest form?

   **F.** $\frac{3}{7}$      **G.** $\frac{1}{2}$      **H.** $\frac{4}{7}$      **I.** $\frac{7}{10}$      2. _____

3. The table shows the animals at a petting zoo. What is the ratio of goats to the total number of animals?

   | Animals at a Petting Zoo | |
   |---|---|
   | Chickens | 2 |
   | Deer | 4 |
   | Donkey | 3 |
   | Goats | 6 |

   **A.** $\frac{5}{2}$      **B.** $\frac{9}{6}$      **C.** $\frac{6}{9}$      **D.** $\frac{2}{5}$      3. _____

**For Exercises 4 and 5, what is each rate written as a unit rate?**

4. 9 pages in 3 hours

   **F.** $\frac{1 \text{ page}}{3 \text{ h}}$      **G.** $\frac{3 \text{ pages}}{1 \text{ h}}$      **H.** $\frac{6 \text{ pages}}{1 \text{ h}}$      **I.** $\frac{9 \text{ pages}}{3 \text{ h}}$      4. _____

5. 20 pounds for $4

   **A.** $\frac{5 \text{ lb}}{\$1}$      **B.** $\frac{4 \text{ lb}}{\$1}$      **C.** $\frac{1 \text{ lb}}{\$4}$      **D.** $\frac{1 \text{ lb}}{\$5}$      5. _____

6. A manatee surfaced for air 2 times in 60 seconds. How many seconds went by before the manatee surfaced the first time if it held its breath for the same rate?

   **F.** 60 s      **G.** 30 s      **H.** 20 s      **I.** 10 s      6. _____

**For Exercises 7 and 8, use the ratio table given to solve each problem.**

7. Nicholas bought 5 pens for $3. At this rate, how much will he spend on 10 pens?

| Number of Pens | 5 | 10 |
|---|---|---|
| Money Spent ($) | 3 | ■ |

   **A.** $10        **B.** $6        **C.** $5        **D.** $3                    7. _____

8. Rashida bought 3 tickets to a concert for $75. At this rate, how much would 5 tickets cost?

| Number of Tickets | 3 | | 5 |
|---|---|---|---|
| Money Spent ($) | 75 | | ■ |

   **F.** $150        **G.** $125        **H.** $100        **I.** $75                    8. _____

9. Gina read 120 pages in 4 days. Which reading rate is equivalent?

   **A.** 75 pages in 3 days        **C.** 105 pages in 5 days
   **B.** 60 pages in 2 days        **D.** 80 pages in 3 days                    9. _____

10. Five notebooks cost $20. At this rate, how much would 4 notebooks cost?

   **F.** $16        **G.** $20        **H.** $25        **I.** $30                    10. _____

11. According to the key of a particular park map, every 5 inches on the map represents 3 miles of trails in the park. How many inches on the map represents 9 miles on the park trails?

   **A.** 15 in.        **B.** 9 in.        **C.** 5 in.        **D.** 3 in.                    11. _____

12. One serving of a fruit punch recipe serves 16 people. Which of the following lists the ordered pairs (number of servings, number of people served) for 1, 2, 3, and 4 servings?

   **F.** (1, 16), (2, 32), (3, 48), (4, 64)

   **G.** (16, 1), (32, 2), (48, 3), (64, 4)

   **H.** (1, 16), (2, 32), (3, 48), (4, 60)

   **I.** (1, 32), (2, 48), (3, 64), (4, 80)                    12. _____

# Test, Form 1B

**Write the letter for the correct answer in the blank at the right of each question.**

1. What is the greatest common factor of 36 and 42?

   **A.** 12          **B.** 8          **C.** 6          **D.** 2          1. _____

2. At a summer camp, there are 50 girls out of 80 campers. What is this ratio written as a fraction in simplest form?

   **F.** $\frac{8}{10}$          **G.** $\frac{5}{8}$          **H.** $\frac{1}{2}$          **I.** $\frac{3}{8}$          2. _____

3. The table lists the number of students from Windy Brook Middle School at the state fair. What is the ratio of sixth graders to the total number of students at the fair?

   | Students at the State Fair | |
   |---|---|
   | 5th graders | 6 |
   | 6th graders | 4 |
   | 7th graders | 5 |
   | 8th graders | 3 |

   **A.** $\frac{9}{2}$          **B.** $\frac{14}{4}$          **C.** $\frac{4}{14}$          **D.** $\frac{2}{9}$          3. _____

**For Excercises 4 and 5, what is each rate written as a unit rate?**

4. 6 miles in 2 hours

   **F.** $\frac{6 \text{ mi}}{1 \text{ h}}$          **G.** $\frac{4 \text{ mi}}{1 \text{ h}}$          **H.** $\frac{3 \text{ mi}}{1 \text{ h}}$          **I.** $\frac{1 \text{ mi}}{3 \text{ h}}$          4. _____

5. 18 pounds for \$9

   **A.** $\frac{2 \text{ lb}}{\$1}$          **B.** $\frac{6 \text{ lb}}{\$1}$          **C.** $\frac{1 \text{ lb}}{\$2}$          **D.** $\frac{1 \text{ lb}}{\$6}$          5. _____

6. A manatee surfaced for air 3 times in 120 seconds. How many seconds went by before the manatee surfaced the first time if it held its breath for the same rate?

   **F.** 10 s          **G.** 20 s          **H.** 30 s          **I.** 40 s          6. _____

# Test, Form 1B    (continued)

**For Exercises 7 and 8, use the ratio table given to solve each problem.**

7. Mr. Chen bought 3 pounds of ham for $21. At this rate, how much would 9 pounds of ham cost?

| Cost ($) | 21 | ■ |
|---|---|---|
| Weight (lb) | 3 | 9 |

   **A.** $9        **B.** $21        **C.** $27        **D.** $63

   7. _____

8. Mr. Paulson has been on a train for 4 hours and traveled 320 miles. At this rate, how far will he have traveled in 5 hours?

| Time (h) | 4 | | 5 |
|---|---|---|---|
| Distance (mi) | 320 | | ■ |

   **F.** 450 mi      **G.** 400 mi      **H.** 320 mi      **I.** 80 mi

   8. _____

9. Juan read 300 pages in 5 days. Which reading rate is equivalent?

   **A.** 150 pages in 3 days        **C.** 105 pages in 1 day

   **B.** 120 pages in 2 days        **D.** 100 pages in 3 days

   9. _____

10. A scanner can scan 40 pages in 4 minutes. At this rate, how many pages can it scan in 2 minutes?

    **F.** 80 pages      **G.** 40 pages      **H.** 20 pages      **I.** 10 pages

    10. _____

11. Chantel counted 48 books on 6 shelves in the library. How many books would she expect to count on 12 shelves?

    **A.** 24 books      **B.** 56 books      **C.** 96 books      **D.** 144 books

    11. _____

12. One package of fruit drink contains 6 drinks. Which of the following lists the ordered pairs (packages, number of drinks) for 1, 2, 3, and 4 packages of fruit drinks?

    **F.** (6, 1), (12, 2), (18, 3), (24, 4)

    **G.** (1, 12), (2, 18), (3, 24), (4, 30)

    **H.** (0, 0), (1, 6), (2, 12), (3, 18)

    **I.** (1, 6), (2, 12), (3, 18), (4, 24)

    12. _____

# Test, Form 2A

**Write the letter for the correct answer in the blank at the right of each question.**

1. One bus leaves a stop every 12 minutes. A second bus leaves the same stop every 18 minutes. If they both leave at 3:20 P.M., at what time will they next leave together?

   **A.** 3:32 P.M.    **B.** 3:38 P.M.    **C.** 3:50 P.M.    **D.** 3:56 P.M.

   1. _____

2. In an orchestra, there are 42 woodwind players to 56 brass players. What is the ratio of woodwinds to brass written as a fraction in simplest form?

   **F.** $\frac{4}{3}$    **G.** $\frac{7}{8}$    **H.** $\frac{6}{8}$    **I.** $\frac{3}{4}$

   2. _____

3. The table shows the number of trees at Citrus Orchards. What is the ratio of orange trees to the total number of trees?

   | Citrus Orchard | |
   |---|---|
   | **Trees** | **Amount of Trees** |
   | Lemon | 30 |
   | Lime | 14 |
   | Orange | 12 |

   **A.** $\frac{14}{3}$    **B.** $\frac{3}{14}$    **C.** $\frac{12}{44}$    **D.** $\frac{3}{11}$

   3. _____

**For Excercises 4 and 5, what is each rate written as a unit rate?**

4. 350 kilometers in 5 hours

   **F.** $\frac{70 \text{ km}}{1 \text{ hr}}$    **G.** $\frac{50 \text{ km}}{1 \text{ hr}}$    **H.** $\frac{1 \text{ km}}{5 \text{ hr}}$    **I.** $\frac{1 \text{ km}}{7 \text{ hr}}$

   4. _____

5. $80 for 16 tickets

   **A.** $\frac{\$16}{1 \text{ ticket}}$    **B.** $\frac{1 \text{ ticket}}{\$16}$    **C.** $\frac{\$5}{1 \text{ ticket}}$    **D.** $\frac{1 \text{ ticket}}{\$5}$

   5. _____

6. It took Jaivin 18 minutes to jog 4 laps. How many minutes did it take to jog each lap at this rate?

   **F.** 2.2 min    **G.** 4.5 min    **H.** 5.4 min    **I.** 9 min

   6. _____

7. Kelly can type 120 words in 3 minutes. How many words can she type in 1 minute at this rate?

   **A.** 30 words    **B.** 40 words    **C.** 50 words    **D.** 60 words

   7. _____

# Test, Form 2A   *(continued)*

**For Exercises 8 and 9, use the ratio table given to solve each problem.**

8. The jumping team can jump 36 times in 9 seconds. At this rate, how many jumps can they make in 27 seconds?

| Jumps | 36 | ■ |
|---|---|---|
| Time (s) | 9 | 27 |

   **F.** 12 jumps   **G.** 36 jumps   **H.** 108 jumps   **I.** 144 jumps

8. _____

9. A customer at a raceway can drive around the track 54 times for $12. At this rate, how many times can the customer drive around the track for $8?

| Number of Times Around Track | 54 | | ■ |
|---|---|---|---|
| Cost ($) | 12 | | 8 |

9. _____

   **A.** 36 times   **B.** 21 times   **C.** 9 times   **D.** 6 times

10. A china shop receives a shipment of 125 dishes in 5 boxes. At this rate, how many dishes will it receive in 15 boxes?

10. _____

11. There are 207 students in the 9 classes at East Middle School. At this rate, how many students are in 6 classes?

11. _____

12. A sewing club used 48 feet of fabric to make 8 blankets. At this rate, how many blankets can be made from 12 yards of fabric?

12. _____

13. George found that he blinked 85 times in 15 minutes. At this rate, how many seconds passed during 51 blinks?

13. _____

14. Liam practiced piano 24 times in 4 weeks. His brother practiced piano 12 times in 14 days. Are the rates at which each brother practiced equivalent? Explain your reasoning.

14. _____

15. Nick can read 3 pages in 1 minute. Write the ordered pairs (number of minutes, number of pages read) for Nick reading 0, 1, 2, and 3 minutes.

15. _____

# Test, Form 2B

SCORE _____

**Write the letter for the correct answer in the blank at the right of each question.**

1. One bus leaves a stop every 9 minutes. A second bus leaves the stop every 15 minutes. If they both leave at 1:25 P.M., at what time will they next leave together?

   **A.** 1:40 P.M.　　**B.** 1:49 P.M.　　**C.** 2:10 P.M.　　**D.** 2:25 P.M.

   1. _____

2. In an art class, there are 32 pens to 40 brushes. What is the ratio of pens to brushes written as a fraction in simplest form?

   **F.** $\frac{4}{5}$　　　**G.** $\frac{2}{5}$　　　　**H.** $\frac{8}{10}$　　　**I.** $\frac{5}{4}$

   2. _____

3. The table shows the number of votes each student received for class president. What is the ratio of votes for Hatim to the total number of votes?

   | Students Class Presidential Votes ||
   | Candidate | Number of Votes |
   | --- | --- |
   | Brown | 15 |
   | Shui | 29 |
   | Hatim | 16 |

   **A.** $\frac{15}{4}$　　　**B.** $\frac{4}{11}$　　　**C.** $\frac{4}{15}$　　　**D.** $\frac{1}{4}$

   3. _____

**For Exercises 4 and 5, what is each rate written as a unit rate?**

4. 180 miles on 6 gallons

   **F.** $\frac{1 \text{ mi}}{3 \text{ gal}}$　　**G.** $\frac{1 \text{ mi}}{9 \text{ gal}}$　　**H.** $\frac{30 \text{ mi}}{1 \text{ gal}}$　　**I.** $\frac{90 \text{ mi}}{1 \text{ gal}}$

   4. _____

5. $120 for 15 books

   **A.** $\frac{\$8}{1 \text{ book}}$　　**B.** $\frac{1 \text{ book}}{\$8}$　　**C.** $\frac{\$15}{1 \text{ book}}$　　**D.** $\frac{1 \text{ book}}{\$15}$

   5. _____

6. It took Cedric 42 minutes to jog 12 laps. At this rate, how many minutes did it take to jog each lap?

   **F.** 2.28 min　　**G.** 3.5 min　　**H.** 4 min　　**I.** 4.5 min

   6. _____

7. Julia can type 150 words in 5 minutes. At this rate, how many words can she type in 1 minute?

   **A.** 30 words　　**B.** 40 words　　**C.** 50 words　　**D.** 60 words

   7. _____

# Test, Form 2B  (continued)

**For Exercises 8 and 9, use the ratio table given to solve each problem.**

8. A hair salon bought 12 headbands for $24. At this rate, how much will 48 headbands cost?

| Headbands | 12 | 48 |
|---|---|---|
| Cost ($) | 24 | ■ |

   **F.** $24      **G.** $48      **H.** $60      **I.** $96      8. _____

9. There are 8 boxes, which contain a total of 56 sweaters, in a store warehouse. If the store needs 35 sweaters, how many boxes should they take?

| Sweaters | 56 | | 35 |
|---|---|---|---|
| Boxes | 8 | | ■ |

   **F.** 4 boxes      **G.** 5 boxes      **H.** 7 boxes      **I.** 9 boxes      9. _____

10. For a class trip, 114 students are put on 4 buses. At this rate, how many students would be on 8 buses?

   10. _____

11. While hiking, Mr. Finds burns 808 Calories in 8 hours. At this rate, how many Calories does he burn in 60 minutes?

   11. _____

12. Margo won 24 out of her last 36 tennis matches. At this rate, predict how many matches she would win out of her next 27 matches?

   12. _____

13. A store manager orders T-shirts so that 15 out of every 35 are a medium. How many medium T-shirts would you expect to find when there are 126 T-shirts on a rack?

   13. _____

14. Mia saves $28 in 8 weeks. Her sister saves $18 in 24 days. Are the rates at which each sister saves equivalent? Explain your reasoning.

   14. _____

15. Luke can read 4 pages in 1 minute. Write the ordered pairs (number of minutes, number of pages read) for Luke reading 0, 1, 2, and 3 minutes.

   15. _____

# Test, Form 3A

1. Carlo attends art class every 4 weeks, chess club every 2 weeks, and fencing lessons every 3 weeks. If he attended all three this week, when will he attend all three again?

1. _____

2. In a certain area, there are 35 houses to 55 businesses. Write the ratio of houses to businesses as a fraction in simplest form. Then explain its meaning.

2. _____

3. The table shows results of a survey about the types of video games students own. Find the ratio of the number of video games rated Everyone 10+ to the total number of video games.

| Ratings Video Games | |
|---|---|
| Rated | Number of Games |
| Early Childhood | 133 |
| Everyone | 130 |
| Everyone 10+ | 140 |
| Teen | 212 |

3. _____

4. On his fruit stand, Mr. Roberts has 13 papayas, 23 star fruits, 35 mangos, and 19 strawberries. What is the ratio of the number of mangos to the total number of pieces of fruit?

4. _____

**Write each ratio as a unit rate.**

5. 162 heartbeats in 60 seconds

5. _____

6. 216 diving students to 12 instructors

6. _____

7. 630 meters in 18 seconds

7. _____

# Test, Form 3A   (continued)

**For Exercises 8 and 9, use the ratio table given to solve each problem.**

8. Maggie's grandmother uses 16 limes to make 2 key lime pies. At this rate, how many limes does she need to make 6 key lime pies?

| Limes | 16 | ■ |
|-------|-----|----|
| Pies | 2 | 6 |

8. _____

9. Ms. Sims traveled to 42 countries in 60 days. At this rate, how many countries would she travel to in 40 days?

| Countries | 42 | ■ |
|-----------|-----|----|
| Days | 60 | 40 |

9. _____

10. After working for 25 hours, Lamar made $375. After working for 40 hours, Lamar made $600. Predict how much he would make after 10 hours of work.

10. _____

11. Determine if the rates $168 raised for washing 24 cars and $280 raised for washing 40 cars are equivalent. Explain your reasoning.

11. _____

12. Suppose 40 out of 90 people are bowlers and 3 out of every 16 of the bowlers have their own bowling ball. At the same rates, in a group of 192 people, predict how many you would expect to have a bowling ball.

12. _____

13. A particular amusement park ride will make 156 revolutions in 180 seconds. At this rate, how many revolutions will it make in 2 minutes?

13. _____

14. The costume designer for the school play uses 4 feet of fabric to make 8 costumes. At the same rate, how many costumes can the designer make with 4 yards of fabric?

14. _____

15. Nina can type 45 words per minute. Write the ordered pairs (number of minutes, number of words) Nina can type for 0, 1, 2, and 3 minutes.

15. _____

# Test, Form 3B

1. Clea attends music lessons every 2 weeks, dance class every 3 weeks, and has a dental appointment every 6 weeks. If she attended all three this week, when will she attend all three again?

1. _____

2. In a certain theater, there are 16 empty seats out of 240 total seats. Write the ratio of empty seats to total seats as a fraction in simplest form. Then explain its meaning.

2. _____

3. Students in a sixth grade class were asked which team was their favorite. The results are shown in the table. Find the ratio of students who preferred the Buccaneers to the total number of responses.

| Favorite Sports Teams | |
|---|---|
| **Teams** | **Number of Students** |
| Tigers | 35 |
| Panthers | 45 |
| Buccaneers | 74 |
| Aviators | 52 |

3. _____

4. On his shelves, Trenton had 15 biographies, 15 science fiction books, 7 reference books, and 11 mystery books. Find the ratio of biographies to the total number of books.

4. _____

**Write each ratio as a unit rate.**

5. 424 kilometers in 8 hours

5. _____

6. $60 for 12 months

6. _____

7. 192 hot dogs in 24 minutes

7. _____

# Test, Form 3B   (continued)

**For Exercises 8 and 9, use the ratio table given to solve each problem.**

8. Daurie rides on her motorcycle for 252 miles on 6 gallons of gas. At this rate, how far will she get on 2 gallons of gas?

| Distance (mi) | ■ | 252 |
|---|---|---|
| Gasoline (gal) | 2 | 6 |

8. _____

9. Casey took 25 tests in 135 days. At this rate, in how many days will she take another 10 tests?

| Days | 135 | | ■ |
|---|---|---|---|
| Tests | 25 | | 10 |

9. _____

10. After working for 24 hours, Zoe made $234. After working for 40 hours, Zoe made $390. Predict how much she would make after 10 hours of work.

10. _____

11. Determine if the rates $24 for 6 hours of work and $36 for 8 hours of work are equivalent. Explain your reasoning.

11. _____

12. Suppose 90 out of 120 people said they like to work out and 5 out of every 18 of the people who like to work out have a gym membership. At the same rates, in a group of 216 people, predict how many you would expect to have a gym membership.

12. _____

13. Mrs. Davis bought 6 pounds of grapes for $18. At this rate, how much would 72 ounces of grapes have cost?

13. _____

14. Jack's plant grew 50 inches in 100 weeks. At this rate, how much did it grow in 210 days?

14. _____

15. Octavia can type 60 words per minute. Write the ordered pairs (number of minutes, number of words) for Octavia typing for 0, 1, 2, and 3 minutes.

15. _____

# Are You Ready?

## Review

### Example

**Find the GCF of 20 and 48.**

First make an organized list of the factors for each number.

20: 1, 2, 4, 5, 10, 20
48: 1, 2, 3, 4, 6, 8, 12, 16, 24, 48

The common factors are 1, 2, and 4, and the greatest of these is 4. So, the greatest common factor, or GCF, of 20 and 48 is 4.

### Exercises

**Find the GCF of each set of numbers.**

1. 22 and 36

1. _____

2. 12 and 18

2. _____

3. 40 and 56

3. _____

4. 28 and 70

4. _____

5. 32 and 76

5. _____

6. 45, 51, and 63

6. _____

7. 24, 60, and 72

7. _____

8. 35, 65, and 90

8. _____

9. 21, 42, and 84

9. _____

10. 18, 27, and 30

10. _____

# Are You Ready?

## Practice

**Find the GCF of each set of numbers.**

**1.** 84 and 108

**2.** 12 and 42

**3.** 28 and 70

**4.** 9, 15, and 63

**5.** 18, 54, and 72

**6.** 36, 80, and 92

**7. SALES** A department store recorded the
amount of money made on sweaters each day.
If the sweaters each cost the same amount,
what is the highest possible cost of each
sweater?

| Sweaters Sold | |
|---|---|
| **Day** | **Money Made** |
| Monday | $66 |
| Tuesday | $110 |
| Wednesday | $132 |

**Find the LCM for each set of numbers.**

**8.** 3 and 11

**9.** 5 and 9

**10.** 6 and 2

**11.** 3, 7, and 9

**12. PATTERNS** Which three common multiples for
3 and 8 are missing from the list below?
24, 48, 72, ■, ■, ■, 168, 192, . . .

**1.** _____

**2.** _____

**3.** _____

**4.** _____

**5.** _____

**6.** _____

**7.** _____

**8.** _____

**9.** _____

**10.** _____

**11.** _____

**12.** _____

# Are You Ready?

## Apply

1. **GARDENING** Raven wants to plant 35 marigold plants, 20 zinnia plants, and 15 daylily plants in her flower garden. If she puts the same number of plants in each row and if each row has only one type of plant, what is the greatest number of plants she can put in one row? Explain.

2. **BAND** The high school jazz band rehearses sometimes with exactly 4 people in every row on stage. Sometimes it rehearses with exactly 5 people in every row. What is the least number of people that can be in the band?

3. **FARMING** Omar is planting a small plot of corn for his family. He has enough seeds to plant 9 or 12 plants in each row. What is the least number of seeds Omar could have?

4. **FIELD TRIP** The sixth-grade teachers collected money from students during three days for theater tickets to a play. The ticket cost a whole number of dollars. If every student paid the same amount, what is the most the tickets could cost per student?

| Monday | $64 |
| Tuesday | $36 |
| Wednesday | $52 |

5. **GIFTS** Yara is making gift baskets for her neighbors. She has 12 chocolate chip cookies, 9 walnut cookies, and 15 sugar cookies to put in the baskets. Each basket must have the same number of cookies in it. Without mixing cookies, what is the greatest number of cookies Yara can put in each basket? Explain.

6. **CLASS PHOTO** The sixth-grade class wants to make a class photo. The students can stand in rows with 7 students in each row, or they can stand with 8 in each row. What is the least number of students that can be in the sixth-grade class?

# Diagnostic Test

**Find the GCF of each set of numbers.**

1. 14 and 63

2. 15 and 70

3. 34 and 51

4. 18, 30, and 66

5. 22, 33, and 88

6. 54, 72, and 81

1. _____

2. _____

3. _____

4. _____

5. _____

6. _____

7. **LAPS** Ernesto recorded the distance he swam in the school swimming pool for a week. If he swam only complete pool lengths, what is the greatest possible length of the pool?

| Distance Swam in School Pool | |
|---|---|
| Day | Distance |
| Monday | 100 meters |
| Tuesday | 150 meters |
| Wednesday | 75 meters |
| Thursday | 125 meters |
| Friday | 175 meters |

7. _____

**Find the LCM for each set of numbers.**

8. 4 and 8

9. 7 and 6

10. 3 and 9

11. 4, 7, and 8

12. 6, 8, and 12

13. 2, 3, and 13

8. _____

9. _____

10. _____

11. _____

12. _____

13. _____

14. **PATTERNS** Which three common multiples for 4 and 9 are missing from the list below?
36, 72, 108, ■, ■, ■, 252, 288, . . .

14. _____

**Course 1 • Chapter 2** Fractions, Decimals, and Percents

# Pretest

**Write each decimal as a fraction in simplest form.**

1. 0.75

2. 0.4

**Write each fraction as a decimal.**

3. $\dfrac{4}{5}$

4. $\dfrac{1}{25}$

**Write each percent as a fraction in simplest form.**

5. 32%

6. 48%

**Write each fraction as a percent.**

7. $\dfrac{7}{20}$

8. $\dfrac{1}{5}$

**Write each percent as a decimal.**

9. 12%

10. 64%

**Write each decimal as a percent.**

11. 0.85

12. 0.73

**Replace each ● with <, >, or = to make a true statement.**

13. $\dfrac{2}{3}$ ● $\dfrac{3}{5}$      14. $\dfrac{1}{8}$ ● $\dfrac{1}{4}$

**Find the percent of each number.**

15. 6% of 38

16. 115% of 212

1. _____

2. _____

3. _____

4. _____

5. _____

6. _____

7. _____

8. _____

9. _____

10. _____

11. _____

12. _____

13. _____

14. _____

15. _____

16. _____

# Chapter Quiz

**Write each fraction or mixed number as a decimal.**

1. $\frac{7}{8}$

2. $\frac{7}{15}$

3. $4\frac{3}{25}$

**Write each decimal as a fraction or mixed number in simplest form.**

4. 0.45

5. 0.125

6. 6.04

7. **HEALTH** The average human body temperature is 98.6°F. Write 98.6 as a mixed number in simplest form.

**Write each percent as a decimal.**

8. 21%

9. 0.4%

**Write each decimal as a percent.**

10. 0.35

11. 0.812

**Write each percent as a fraction in simplest form.**

12. $12\frac{1}{2}\%$

13. 48%

**Write each fraction as a percent.**

14. $\frac{7}{20}$

15. $\frac{3}{5}$

**Write each percent as a decimal and as a mixed number or fraction in simplest form.**

16. 300%

17. 0.5%

**Write each decimal as a percent.**

18. 7.5

19. 0.068

20. **FAMILY** Marcos talks to his grandfather in Alabama for about 95 minutes per month. About how long does he talk each day, on the average? (*Hint*: Use 1 month ≈ 30 days.)

1. _____

2. _____

3. _____

4. _____

5. _____

6. _____

7. _____

8. _____

9. _____

10. _____

11. _____

12. _____

13. _____

14. _____

15. _____

16. _____

17. _____

18. _____

19. _____

20. _____

# Vocabulary Test

| least common denominator (LCD) | percent<br>percent proportion | proportion<br>rational number |
|---|---|---|

**Choose from the terms above to complete each sentence.**

1. For $\frac{3}{4}$ and $\frac{2}{3}$, the _____ is 12.

    1. _____

2. $\frac{4}{5}$ is a _____ because it can be written as a fraction.

    2. _____

3. To write 0.64 as a _____, multiply it by 100.

    3. _____

4. A _____ is an equation that shows that two ratios are equivalent.

    4. _____

5. A _____ is a ratio that compares a number to 100.

    5. _____

6. In a _____, one ratio compares a part to the whole. The other ratio is the equivalent percent written as a fraction with a denominator of 100.

    6. _____

**Define each term in your own words.**

7. least common denominator (LCD)

    7. _____

8. rational number

    8. _____

9. percent

    9. _____

# Standardized Test Practice

**Read each question. Then fill in the correct answer on the answer sheet provided by your teacher or on a sheet of paper.**

1. At Medina Middle School, 2% of the students ride bikes to school. If 950 students attend Medina Middle School, how many students ride a bike to school?

| Transportation to School | Percent of Students |
|---|---|
| Ride a bus | 73 |
| Ride a bike | 2 |
| Walk | 7 |
| Parents drive | 18 |

   A. 17
   B. 18
   C. 19
   D. 20

2. The formula $C = \frac{5}{9}(F - 32)$ can be used to convert a temperature from degrees Fahrenheit to degrees Celsius. Which of the following is closet in value to $\frac{5}{9}$?
   F. 5.9
   G. 4
   H. 1.8
   I. 0.56

3. Sanford's Shoe Store received a shipment of shoes for its newest location. The manager determined that 35% of the shoes were athletic shoes. What fraction of the shoes were athletic shoes?
   A. $\frac{7}{20}$
   B. $\frac{1}{6}$
   C. $\frac{3}{8}$
   D. $\frac{13}{20}$

4. ▰▱ **GRIDDED RESPONSE** Mark got 18 out of 25 questions correct on his science test. What percent of the questions did he get correct?

5. The graph shows the elements found in Earth's crust. What fraction of Earth's crust is silicon?

**Elements in Earth's Crust**

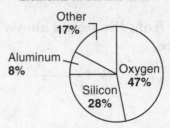

   F. $\frac{1}{28}$
   G. $\frac{7}{25}$
   H. $\frac{2}{5}$
   I. $\frac{1}{2}$

6. The fastest fish in the world is the sailfish. If a sailfish could maintain its speed, as shown in the table, how many miles could the sailfish travel in 6 hours?

| Hours Traveled | 0 | 1 | 2 | 3 | 4 | 5 | 6 |
|---|---|---|---|---|---|---|---|
| Miles Traveled | 0 | 68 | 136 | 204 | 272 | 340 | ? |

   A. 6 miles
   B. 68 miles
   C. 408 miles
   D. 476 miles

7. ▰▱ **GRIDDED RESPONSE** Carlita planted 4 flowers in 9 minutes. About how many flowers can Carlita plant in 36 minutes?

8. What is the ratio of people to buses?

| 6 buses | 150 people |
|---|---|

   F. 1:25
   G. 25:1
   H. 6:150
   I. 156:6

9. A school population was predicted to increase by 50 students a year for the next 10 years. If the current population is 700 students, what will the enrollment be after 10 years?

   A. 50

   B. 500

   C. 1,200

   D. 7,000

10. Pam is painting a total of 45 square feet in her bedroom. If she has painted 60% so far, how many square feet of the bedroom does Pam still need to paint?

   F. 30

   G. 27

   H. 18

   I. 10

11. A class of 26 students ordered the items shown in the table.

   | Quantity | Item | Unit Price |
   |----------|------|------------|
   | 27 | Doughnut | $1.25 |
   | 5 | Gallon of orange juice | $1.99 |
   | 1 | Package of napkins | $1.50 |

   If the class agreed to split the cost evenly, about how much will each student pay?

   A. $3.00

   B. $2.50

   C. $1.50

   D. $1.00

12. SHORT RESPONSE Refer to the table below.

   | Williams Family Budget | |
   |------------------------|------|
   | House expenses | 43% |
   | Transportation | 16% |
   | Clothing and entertainment | 12% |
   | Food | 11% |
   | Savings account | 10% |
   | Other | 8% |

   If the Williams family has a monthly income of $3,000, how much do they spend on clothing and entertainment?

13. EXTENDED RESPONSE Copy the models below. Both models have the same area.

   **Model A**

   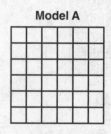

   **Model B**

   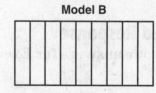

   *Part A*  Shade 0.25 of Model A.

   *Part B*  Shade $\frac{1}{3}$ of Model B.

   *Part C*  Which model has the greater fraction of shaded area? Explain your answer.

# Student Recording Sheet

*Use this recording sheet with the Standardized Test Practice pages.*

**Fill in the correct answer. For gridded-response questions, write your answers in the boxes on the answer grid and fill in the bubbles to match your answers.**

1. Ⓐ Ⓑ Ⓒ Ⓓ

2. Ⓕ Ⓖ Ⓗ Ⓘ

3. Ⓐ Ⓑ Ⓒ Ⓓ

4.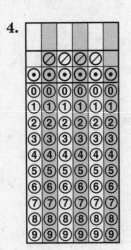

5. Ⓕ Ⓖ Ⓗ Ⓘ

6. Ⓐ Ⓑ Ⓒ Ⓓ

7.

8. Ⓕ Ⓖ Ⓗ Ⓘ

9. Ⓐ Ⓑ Ⓒ Ⓓ

10. Ⓕ Ⓖ Ⓗ Ⓘ

11. Ⓐ Ⓑ Ⓒ Ⓓ

12. _____

## Extended Response

Record your answers for Exercise 13 on the back of this paper.

# Extended-Response Test

Demonstrate your knowledge by giving a clear, concise solution to each problem. Be sure to include all relevant drawings and justify your answers. You may show your solution in more than one way or investigate beyond the requirements of the problem. If necessary, record your answer on another piece of paper.

1. A store is marking down clothing 30% for the back-to-school sale.

   **a.** Tell how to write a percent as a fraction and as a decimal.

   **b.** Estimate the amount taken off of a jacket regularly priced $44. Explain your reasoning.

2. Use the table that shows average winter monthly rainfall.

| Portion of Month with Precipitation | | |
|---|---|---|
| Month | River City | Lakeview |
| November | $\frac{3}{8}$ | 40% |
| December | $\frac{1}{8}$ | 32% |
| January | $\frac{1}{2}$ | 38% |

   **a.** Write River City's rainfall as a percent for each month. Show your work.

   **b.** Write Lakeview's rainfall for November and December as decimals. Compare each month to River City's rainfall for November and December.

3. Tell how to write a fraction as a percent.

# Extended-Response Rubric

| Score | Description |
|-------|-------------|
| 4 | A score of four is a response in which the student demonstrates a thorough understanding of the mathematics concepts and/or procedures embodied in the task. The student has responded correctly to the task, used mathematically sound procedures, and provided clear and complete explanations and interpretations. <br><br> The response may contain minor flaws that do not detract from the demonstration of a thorough understanding. |
| 3 | A score of three is a response in which the student demonstrates an understanding of the mathematics concepts and/or procedures embodied in the task. The student's response to the task is essentially correct with the mathematical procedures used and the explanations and interpretations provided demonstrating an essential but less than thorough understanding. <br><br> The response may contain minor flaws that reflect inattentive execution of mathematical procedures or indications of some misunderstanding of the underlying mathematics concepts and/or procedures. |
| 2 | A score of two indicates that the student has demonstrated only a partial understanding of the mathematics concepts and/or procedures embodied in the task. Although the student may have used the correct approach to obtaining a solution or may have provided a correct solution, the student's work lacks an essential understanding of the underlying mathematical concepts. <br><br> The response contains errors related to misunderstanding important aspects of the task, misuse of mathematical procedures, or faulty interpretations of results. |
| 1 | A score of one indicates that the student has demonstrated a very limited understanding of the mathematics concepts and/or procedures embodied in the task. The student's response is incomplete and exhibits many flaws. Although the student's response has addressed some of the conditions of the task, the student reached an inadequate conclusion and/or provided reasoning that was faulty or incomplete. <br><br> The response exhibits many flaws or may be incomplete. |
| 0 | A score of zero indicates that the student has provided no response at all, or a completely incorrect or uninterpretable response, or demonstrated insufficient understanding of the mathematics concepts and/or procedures embodied in the task. For example, a student may provide some work that is mathematically correct, but the work does not demonstrate even a rudimentary understanding of the primary focus of the task. |

# Test, Form 1A

SCORE _____

**Write the letter for the correct answer in the blank at the right of each question.**

1. What is $\frac{1}{4}$ written as a percent?
   **A.** 14%   **B.** 25%   **C.** 30%   **D.** 40%

   1. _____

2. What is 22.5% written as a fraction in simplest form?
   **F.** $\frac{225}{10}$   **G.** $\frac{45}{100}$   **H.** $\frac{9}{40}$   **I.** $\frac{9}{100}$

   2. _____

3. What is 575% written as a mixed number?
   **A.** $5\frac{3}{4}$   **B.** $50\frac{1}{4}$   **C.** $57\frac{1}{2}$   **D.** $500\frac{3}{4}$

   3. _____

4. What is 0.399 written as a percent?
   **F.** 399%   **G.** 39.9%   **H.** 3.99%   **I.** 0.00399%

   4. _____

5. What is $7\frac{1}{5}$ written as a percent?
   **A.** 7.2%   **B.** 27%   **C.** 72%   **D.** 720%

   5. _____

6. What is 86% written as a decimal?
   **F.** 86   **G.** 8.6   **H.** 0.86   **I.** 0.086

   6. _____

7. What is 32% of 60?
   **A.** 19.2   **B.** 32   **C.** 187.5   **D.** 192

   7. _____

8. Find 340% of 35.
   **F.** 0.103   **G.** 70   **H.** 119   **I.** 340

   8. _____

9. Find 0.15% of 43.
   **A.** 64.5   **B.** 6.45   **C.** 0.645   **D.** 0.0645

   9. _____

10. What is 20% of 141?
    **F.** 2.82   **G.** 20   **H.** 28.2   **I.** 70.5

    10. _____

# Test, Form 1A   (continued)

**11.** Which fraction is less than $\frac{1}{2}$?

   **A.** $\frac{3}{8}$       **B.** $\frac{5}{8}$       **C.** $\frac{5}{7}$       **D.** $\frac{9}{16}$       **11.** _____

**12.** Which of the following orders 75%, $\frac{1}{2}$, 0.375, and $\frac{5}{8}$ from least to greatest?

   **F.** $\frac{1}{2}$, 75%, 0.375, $\frac{5}{8}$       **H.** 0.375, $\frac{1}{2}$, 75%, $\frac{5}{8}$

   **G.** $\frac{1}{2}$, 0.375, 75%, $\frac{5}{8}$       **I.** 0.375, $\frac{1}{2}$, $\frac{5}{8}$, 75%       **12.** _____

**13.** What is $6\frac{1}{4}$ written as a decimal?

   **A.** 6.5       **B.** 6.2       **C.** 6.25       **D.** 6.025       **13.** _____

**14.** What is 0.9 written as a fraction in simplest form?

   **F.** $\frac{90}{100}$       **G.** $\frac{9}{10}$       **H.** $\frac{1}{9}$       **I.** $\frac{9}{100}$       **14.** _____

**15.** 16 is 20% of what number?

   **A.** 40       **B.** 80       **C.** 96       **D.** 125       **15.** _____

**16.** A store is having a sale where school supplies are 30% off their original price. A backpack is on sale for $11.20. What was the original price of the backpack?

   **F.** $11.50       **G.** $12       **H.** $14.20       **I.** $16       **16.** _____

**17.** A recipe for guacamole uses 12% lime juice. If a batch contains 0.75 cup of lime juice, how large is the batch of guacamole?

   **A.** 4.5 cups       **B.** 6.25 cups       **C.** 7.5 cups       **D.** 9 cups       **17.** _____

**18.** Fred completed 10 math problems. This is 40% of the number of math problems he has to do. How many math problems must he do?

   **F.** 4       **G.** 17       **H.** 25       **I.** 400       **18.** _____

**Which is the best estimate of each percent?**

**19.** 25% of 395

   **A.** 10       **B.** 100       **C.** 1,000       **D.** 1,600       **19.** _____

**20.** 48% of 60

   **F.** 3,000       **G.** 300       **H.** 30       **I.** 3       **20.** _____

# Test, Form 1B

**Write the letter for the correct answer in the blank at the right of each question.**

1. What is the fraction $\frac{17}{25}$ written as a percent?
   **A.** 17%  **B.** 68%  **C.** 77%  **D.** 680%

   1. _____

2. What is 0.02% written as a decimal?
   **F.** 0.0002  **G.** 0.002  **H.** 0.02  **I.** 2.0

   2. _____

3. What is 39.5% written as a fraction in simplest form?
   **A.** $\frac{395}{10}$  **B.** $\frac{79}{100}$  **C.** $\frac{158}{200}$  **D.** $\frac{79}{200}$

   3. _____

4. What is 24 written as a percent?
   **F.** 2,400%  **G.** 240%  **H.** 2.4%  **I.** 0.24%

   4. _____

5. What is 17% written as a decimal?
   **A.** 0.0017  **B.** 0.017  **C.** 0.17  **D.** 1.7

   5. _____

6. What is 0.02 written as a percent?
   **F.** 0.0002%  **G.** 0.02%  **H.** 2%  **I.** 20%

   6. _____

7. Find 0.45% of 80.
   **A.** 36  **B.** 3.6  **C.** 0.36  **D.** 0.036

   7. _____

8. Find 230% of 46.
   **F.** 105.8  **G.** 10.58  **H.** 1.058  **I.** 0.1058

   8. _____

9. What is 18% of 95?
   **A.** 17.1  **B.** 30.4  **C.** 77.9  **D.** 113

   9. _____

10. What is 65% of 220?
    **F.** 65  **G.** 143  **H.** 155  **I.** 338.5

    10. _____

# Test, Form 1B    *(continued)*

11. Which fraction is greater than $\frac{1}{2}$ and less than $\frac{3}{4}$?

    **A.** $\frac{11}{12}$      **B.** $\frac{7}{8}$      **C.** $\frac{1}{3}$      **D.** $\frac{2}{3}$

    11. _____

12. Which of the following orders 75%, $\frac{1}{6}$, $\frac{3}{8}$, and 0.625 from least to greatest?

    **F.** $\frac{1}{6}$, 75%, $\frac{3}{8}$, 0.625         **G.** $\frac{1}{6}$, $\frac{3}{8}$, 75%, 0.625

    **H.** $\frac{3}{8}$, $\frac{1}{6}$, 0.625, 75%         **I.** $\frac{1}{6}$, $\frac{3}{8}$, 0.625, 75%

    12. _____

13. What is 12.08 written as a mixed number?

    **A.** $12\frac{8}{10}$      **B.** $12\frac{4}{5}$      **C.** $12\frac{4}{25}$      **D.** $12\frac{2}{25}$

    13. _____

14. Beth has made 12 cupcakes. This is 30% of the cupcakes she must make. How many cupcakes must she make?

    **F.** 4      **G.** 17      **H.** 40      **I.** 250

    14. _____

15. What is $\frac{3}{8}$ written as a decimal?

    **A.** 0.35      **B.** 0.275      **C.** 0.375      **D.** 0.25

    15. _____

16. A store is having a sale where movies are 15% off their original price. A movie is on sale for $22.10. What was the original price of the movie?

    **F.** $22.35      **G.** $23      **H.** $26      **I.** $28

    16. _____

17. A recipe for salsa uses 65% tomatoes. If a batch contains 5.2 cups of tomatoes, how large is the batch of salsa?

    **A.** 5.85 cups    **B.** 6 cups    **C.** 7 cups    **D.** 8 cups

    17. _____

18. James completed 18 math problems. This is 30% of the problems he has to do. How many math problems must he do?

    **F.** 48      **G.** 60      **H.** 78      **I.** 540

    18. _____

**Which is the best estimate of each percent?**

19. 49% of 598

    **A.** 3      **B.** 30      **C.** 300      **D.** 3,000

    19. _____

20. 21% of 387

    **F.** 80      **G.** 0.8      **H.** 800      **I.** 8

    20. _____

# Test, Form 2A

**Write the letter for the correct answer in the blank at the right of each question.**

1. What is 2.6% written as a decimal?

   **A.** 0.026     **B.** 0.26     **C.** 26     **D.** 260

   1. _____

2. The library surveyed 240 people about their favorite type of movie. If 15% of the people chose documentaries, how many people chose documentaries?

   **F.** 12 people    **G.** 36 people    **H.** 60 people    **I.** 1600 people

   2. _____

3. The original price of a jacket is $68. The sale price is 30% off the original price. What is the amount off the original price?

   **A.** $20.40     **B.** $30     **C.** $47.60     **D.** $226.67

   3. _____

4. Which of the following orders $\frac{2}{3}$, 25%, $\frac{3}{5}$, and 0.40 from least to greatest?

   **F.** 0.40, $\frac{3}{5}$, 25%, $\frac{2}{3}$        **G.** 25%, 0.40, $\frac{3}{5}$, $\frac{2}{3}$

   **H.** $\frac{2}{3}$, 25%, $\frac{3}{5}$, 0.40        **I.** $\frac{3}{5}$, 0.40, $\frac{2}{3}$, 25%

   4. _____

5. What is 15.08 written as a mixed number?

   **A.** $15\frac{18}{100}$     **B.** $15\frac{2}{25}$     **C.** $15\frac{2}{5}$     **D.** $15\frac{2}{50}$

   5. _____

6. Trey has answered 15 E-mails. This is 20% of E-mails he must answer. How many E-mails does he need to answer?

   **F.** 100     **G.** 75     **H.** 25     **I.** 5

   6. _____

7. A basketball player made 40% of the shots she attempted. If she made 32 baskets, how many shots did she attempt?

   **A.** 32     **B.** 40     **C.** 64     **D.** 80

   7. _____

8. What is $3\frac{9}{20}$ written as a decimal?

   **F.** 3.045     **G.** 3.25     **H.** 3.45     **I.** 3.025

   8. _____

**Which is the best estimate of each percent?**

9. 32% of 19

   **A.** 0.6     **B.** 6     **C.** 60     **D.** 600

   9. _____

10. 48% of 199

   **F.** 0.1     **G.** 1     **H.** 10     **I.** 100

   10. _____

# Test, Form 2A    (continued)

11. Elsa got an 86% on her first science test, a $\frac{47}{50}$ on her second test, and a 0.89 on her third test. Which of the following orders her scores from least to greatest?

 A. 0.86, 0.94, 0.89    C. 0.86, 0.89, 0.94

 B. 0.94, 0.89, 0.84    D. 0.89, 0.86, 0.94

11. _____

12. Cedric can swim a lap in $23\frac{3}{5}$ seconds. Jason can swim a lap in 21.4 seconds. How much faster can Jason swim a lap than Cedric?

 F. 3.1 seconds   G. 2.9 seconds   H. 2.2 seconds   I. 1.9 seconds

12. _____

13. What is 7.5% written as a fraction in simplest form?

 A. $\frac{30}{40}$    B. $\frac{3}{400}$    C. $\frac{3}{4}$    D. $\frac{3}{40}$

13. _____

14. What is 0.15% written as a decimal?

 F. 0.0015    G. 0.015    H. 1.5    I. 15

14. _____

15. What is 345 written as a percent?

 A. 34,500%    B. 3,450%    C. 3.45%    D. 0.345%

15. _____

16. Air contains several gasses in addition to oxygen. Argon makes up 0.0093 of air. What is 0.0093 written as a percent?

 F. 0.0093%    G. 0.93%    H. 9.3%    I. 93%

16. _____

**Find the percent of each number.**

17. 20% of 140              18. 60% of 250

17. _____

18. _____

**Write each fraction as a percent.**

19. $\frac{37}{50}$              20. $\frac{3}{25}$

19. _____

20. _____

**Write each percent as a decimal and as a mixed number or fraction in simplest form.**

21. 185%              22. 0.35%

21. _____

22. _____

23. During a basketball game, Jorell attempted 40 shots and made 18. He says he made 40% of the shots he took. Is Jorell correct? Explain your reasoning.

23. _____

24. The original price of a DVD is $9. The sale price is 20% off the original price. What is the sale price of the DVD?

24. _____

# Test, Form 2B

**Write the letter for the correct answer in the blank at the right of each question.**

1. What is 3.8% written as a decimal?

   **A.** 0.038     **B.** 0.38     **C.** 38     **D.** 380

   1. _____

2. The library surveyed 180 people about their favorite type of movie. If 40% of the people chose action, how many people chose action?

   **F.** 180 people    **G.** 140 people    **H.** 72 people    **I.** 40 people

   2. _____

3. The original price of a pair of jeans is $32. The sale price is 15% off the original price. What is the amount off the original price?

   **A.** $1.75     **B.** $4.80     **C.** $5.25     **D.** $15

   3. _____

4. What is the order of 75%, $\frac{1}{3}$, $\frac{3}{8}$, and 0.625 from least to greatest?

   **F.** $\frac{1}{3}$, 75%, $\frac{3}{8}$, 0.625        **G.** $\frac{3}{8}$, $\frac{1}{3}$, 75%, 0.625

   **H.** $\frac{3}{8}$, $\frac{1}{3}$, 0.625, 75%       **I.** $\frac{1}{3}$, $\frac{3}{8}$, 0.625, 75%

   4. _____

5. What is 14.06 written as a mixed number?

   **A.** $14\frac{6}{10}$     **B.** $14\frac{3}{5}$     **C.** $14\frac{3}{25}$     **D.** $14\frac{3}{50}$

   5. _____

6. Jim has answered 7 E-mails. This is 35% of the E-mails he must answer. How many E-mails does he need to answer?

   **F.** 20     **G.** 13     **H.** 10     **I.** 5

   6. _____

7. In a load of laundry, 60% of the socks are white. If there are 25 socks in the load, how many are white?

   **A.** 12 socks    **B.** 15 socks    **C.** 20 socks    **D.** 25 socks

   7. _____

8. What is $4\frac{7}{20}$ written as a decimal?

   **F.** 4.28     **G.** 4.35     **H.** 4.7     **I.** 4.45

   8. _____

**Which is the best estimate of each percent?**

9. 24% of 398

   **A.** 10     **B.** 100     **C.** 1,000     **D.** 0.1

   9. _____

10. 48% of 159

   **F.** 80     **G.** 8     **H.** 0.8     **I.** 8,000

   10. _____

# Test, Form 2B    (continued)

11. Jeff got a $\frac{23}{25}$ on his first math test, a 93% on his second test, and a
0.89 on his third test. Which of the following orders his scores from
least to greatest?
A. 0.92, 0.93, 0.89          C. 0.89, 0.93, 0.92
B. 0.93, 0.92, 0.89          D. 0.89, 0.92, 0.93

11. _____

12. Jessica can swim a lap in $29\frac{2}{5}$ seconds. Marta can swim a lap in
28.7 seconds. How much faster can Marta swim a lap than Jessica?
F. 1.7 seconds  G. 1.5 seconds  H. 0.7 second  I. 0.5 seconds

12. _____

13. What is 12.5% written as a fraction in simplest form?
A. $\frac{125}{1,000}$     B. $\frac{1}{8}$     C. $\frac{3}{25}$     D. $\frac{5}{4}$

13. _____

14. What is 0.12% written as a decimal?
F. 0.0012     G. 1.2     H. 12     I. 120

14. _____

15. What is 255 written as a percent?
A. 2.55%     B. 25.5%     C. 2,550%     D. 25,500%

15. _____

16. In addition to oxygen, several gasses are in air. Carbon dioxide
makes up 0.00039 of air. What is 0.00039 written as a percent?
F. 39%     G. 0.39%     H. 0.039%     I. 0.00039%

16. _____

**Find the percent of each number.**

17. 17% of 155                    18. 24% of 780

17. _____

**Write each fraction as a percent.**

19. $\frac{47}{50}$                            20. $\frac{4}{25}$

18. _____

19. _____

20. _____

**Write each percent as a decimal and as a mixed number or
fraction in simplest form.**

21. 165%

21. _____

22. 0.65%

22. _____

23. During a basketball game, Dallen attempted 20 shots and made
7. He says he made 35% of the shots he took. Is Dallen correct?
Explain your reasoning.

23. _____

24. The original price of a DVD is $11. The sale price is 30% off
the original price. What is the sale price of the DVD?

24. _____

# Test, Form 3A

1. Sophia downloaded 592 vacation pictures. Of the pictures, 63% show the beach. About how many pictures show the beach? Use a rate per 100 to estimate.

   1. _____

2. Twenty-one students in Michael's classroom are wearing jeans. There are 25 students in his class. Michael says that 80% of his class is wearing jeans. Is Michael correct? Explain your reasoning.

   2. _____

3. During a winter storm, the price of road salt had a 230% increase. Write 230% as a mixed number in simplest form.

   3. _____

4. In a sandwich shop, 22% of the sandwiches they sell are ham. What is 22% written as a decimal?

   4. _____

5. Jana has $5. She wants to buy a notebook for $4.50. The sales tax is 7%. Does she have enough money? Explain your reasoning.

   5. _____

6. The original price of a sweater is $36. The sale price is 85% of the original price. What is the sale price of the sweater?

   6. _____

**Write each percent as a fraction in simplest form.**

7. 5.5%                    8. $87\frac{1}{2}\%$

   7. _____

   8. _____

**Write each fraction as a percent.**

9. $\frac{14}{40}$                    10. $\frac{7}{10}$

   9. _____

   10. _____

**Write each percent as a decimal and as a mixed number or fraction in simplest form.**

11. 220%                   12. 0.08%

   11. _____

   12. _____

**Find the percent of each number.**

13. 34% of 354            14. 92% of 85

   13. _____

   14. _____

**What is the best estimate of each percent?**

15. 20% of 516            16. 24% of 80

   15. _____

   16. _____

# Test, Form 3A (continued)

**For Exercises 17–19, refer to the table.**

17. Do more kids say they want to see into the future or become invisible?

| Superpower | Fraction of Kids |
|---|---|
| become invisible | $\frac{9}{50}$ |
| fly | $\frac{21}{100}$ |
| read minds | $\frac{7}{25}$ |
| see at night | $\frac{1}{25}$ |
| see into the future | $\frac{1}{5}$ |

18. Among those surveyed, which superpower is chosen most often?

19. Which superpower is chosen least often?

17. _____

18. _____

19. _____

20. Eva has $10 for lunch. Her meal costs $8.60. She wants to leave a 20% tip. Does she have enough money? Explain your reasoning.

20. _____

**Write each decimal as a percent.**

21. 0.275          22. 0.08

21. _____

22. _____

**For Exercises 23–25, write each fraction or mixed number as a decimal.**

23. $\frac{11}{20}$          24. $\frac{6}{30}$          25. $8\frac{3}{8}$

23. _____

24. _____

25. _____

26. Elbert made 48% of the shots he took during the basketball game. If he attempted 25 shots, how many times did he score?

26. _____

27. The original price of a DVD is $12. The sale price is 70% of the original price. Find the sale price of the DVD.

27. _____

28. The table shows the wins of three local baseball teams. They all played the same number of games. List the teams in order of the fraction of games won from least to greatest.

| Team | Wins |
|---|---|
| Bears | 57% |
| Tigers | $\frac{5}{8}$ |
| Mustangs | 0.65 |

28. _____

# Test, Form 3B

1. Jason downloaded 288 pictures from his camera. Of the pictures, 38% are of his family. About how many pictures are of his family? Use a rate per 100 to estimate.

1. _____

2. Nine students in Maddie's classroom are wearing red. There are 30 students in her class. Maddie says that 30% of her class is wearing red. Is Maddie correct? Explain your reasoning.

2. _____

3. After an ad campaign, a juice drink's popularity had a 125% increase. Write 125% as a mixed number in simplest form.

3. _____

4. On one day in an ice cream store, 48% of the cones they sell are chocolate. What is 48% written as a decimal?

4. _____

5. Jay has $8. He wants to buy a book for $7.50. The sales tax is 6%. Does he have enough money? Explain your reasoning.

5. _____

6. The original price of a bicycle is $129. The sale price is 80% of the original price. What is the sale price of the bicycle?

6. _____

**Write each percent as a fraction in simplest form.**

7. 7.5%                                   8. $62\frac{1}{2}\%$

7. _____

8. _____

**Write each fraction as a percent.**

9. $\frac{12}{50}$                              10. $\frac{5}{20}$

9. _____

10. _____

**Write each percent as a decimal and as a mixed number or fraction in simplest form.**

11. 345%                                  12. 0.06%

11. _____

**Find the percent of each number.**

12. _____

13. 78% of 160                          14. 12% of 325

13. _____

14. _____

**What is the best estimate of each percent?**

15. 48% of 170                          16. 25% of 320

15. _____

16. _____

# Test, Form 3B  (continued)

**FOR EXERCISES 17–19,
REFER TO THE TABLE.**

| Superpower | Fraction of Kids |
|---|---|
| become invisible | $\frac{1}{25}$ |
| fly | $\frac{1}{5}$ |
| read minds | $\frac{7}{25}$ |
| see at night | $\frac{8}{50}$ |
| see into the future | $\frac{23}{100}$ |

17. Do more kids say they want to see into the future or become invisible?

17. _____

18. Among those surveyed, which superpower is chosen most often?

18. _____

19. Which superpower is chosen least often?

19. _____

20. Dominic has $15 for dinner. His meal costs $13.90. He wants to leave an 18% tip. Does he have enough money? Explain your reasoning.

20. _____

**Write each decimal as a fraction or mixed number in simplest form.**

21. 0.325

22. 0.02

21. _____

22. _____

**For Exercises 23–25, write each fraction or mixed number as a decimal.**

23. $\frac{7}{25}$

24. $\frac{5}{8}$

25. $8\frac{9}{10}$

23. _____

24. _____

25. _____

26. Autry made 45% of the shots he took during the basketball game. If he attempted 20 shots, how many times did he score?

26. _____

27. The original price of a DVD is $12. The sale price is 75% of the original price. Find the sale price of the DVD.

27. _____

28. The table shows the wins of three local baseball teams. They all played the same number of games. List the teams in order of the fraction of games won from least to greatest.

| Team | Wins |
|---|---|
| Bears | 86% |
| Tigers | $\frac{7}{8}$ |
| Mustangs | 0.84 |

28. _____

# Are You Ready?

## Review

### Example 1
**Find 42 ÷ 6.**

$$
\begin{array}{r}
7 \\
6\overline{)42} \\
-42 \\
\hline
0
\end{array}
$$

THINK: What number times 6 is 42?

So, 42 ÷ 6 = 7.

### Example 2
**Find 24 ÷ 2.**

$$
\begin{array}{r}
12 \\
2\overline{)24} \\
-24 \\
\hline
0
\end{array}
$$

THINK: What number times 2 is 24?

So, 24 ÷ 2 = 12.

## Divide.

1. 64 ÷ 8

2. 63 ÷ 7

3. 16 ÷ 4

4. 81 ÷ 9

5. 25 ÷ 5

6. 26 ÷ 13

7. 100 ÷ 10

8. 121 ÷ 11

9. 108 ÷ 12

10. 144 ÷ 12

1. _____

2. _____

3. _____

4. _____

5. _____

6. _____

7. _____

8. _____

9. _____

10. _____

# Are You Ready?

## Practice

**Multiply.**

1. $62 \times 23$

2. $14 \times 31$

3. $28 \times 15$

4. $17 \times 40$

5. $86 \times 20$

6. $39 \times 11$

7. **PAINTING** Mari painted 3 rooms in her house. She spent $52 on paint for each room. How much did she spend on paint?

8. **MUSIC** A store sold 15 CDs for $13 each. What was the total value of the CDs?

9. **FENCES** A farmer put up 234 feet of fence each day for 4 days. How many feet of fence did he put up in all?

**Divide.**

10. $308 \div 4$

11. $488 \div 8$

12. $966 \div 6$

13. $600 \div 3$

14. **WORK** James worked 112 hours in 8 weeks. He worked the same amount of time every week. How many hours did James work each week?

1. _____

2. _____

3. _____

4. _____

5. _____

6. _____

7. _____

8. _____

9. _____

10. _____

11. _____

12. _____

13. _____

14. _____

# Are You Ready?

## Apply

---

**1. CLOTHES** Ellen buys 6 new shirts for $22 each. How much does she spend on shirts?

**2. SHOPPING** Fruit is sold by the bushel at the farmer's market. Andrew buys 12 bushels of peaches to make preserves for the school bake sale. How much does he spend on peaches?

| Apricots | $15 per bushel |
|----------|----------------|
| Peaches  | $13 per bushel |
| Pears    | $12 per bushel |

---

**3. ZOO** The bears at the zoo eat 875 pounds of food each week. How much do they eat per day?

**4. DISTANCE** Mrs. Mendez drives 34 miles each day to take her children to school and run errands. How many miles did she drive in 13 days?

---

**5. RUNNING** The school track team ran 96 miles in 12 days. How many miles did they run per day?

**6. GRASS** Ernie mows grass to make money. He made $324 for mowing 4 lawns last week. If he made the same amount on each lawn, how much did he get paid for each?

---

# Diagnostic Test

**Multiply.**

1. $25 \times 30$

1. _____

2. $41 \times 13$

2. _____

3. $32 \times 17$

3. _____

4. $11 \times 56$

4. _____

5. $93 \times 40$

5. _____

6. $67 \times 33$

6. _____

7. **SHOPPING** Aaron bought two new parkas for $68 each. How much did he spend on the parkas?

7. _____

8. **CHARITY** Tim's sponsors agreed to donate $11 for each mile he ran in a charity race. Tim ran 12 miles. How much did he raise for charity?

8. _____

9. **REMODELING** Adrian put down 325 square feet of carpet each day for 5 days. How many square feet of carpet did he install in all?

9. _____

**Divide.**

10. $392 \div 7$

10. _____

11. $243 \div 9$

11. _____

12. $429 \div 3$

12. _____

13. $693 \div 7$

13. _____

14. **TRAVEL** Viranda's family traveled 585 miles in 9 hours of driving. How many miles did they drive per hour on the average?

14. _____

# Pretest

**Estimate each product.**

1. $3.8 \times 4$

2. $26.5 \times 12$

**Multiply.**

3. $5.6 \times 8$

4. $9 \times 7.03$

5. $3.14 \times 0.06$

6. $6.8 \times 1.5$

**Estimate each quotient.**

7. $63.7 \div 9.1$

8. $31.2 \overline{)93.6}$

**Divide.**

9. $8.1 \div 3$

10. $4.9 \div 7$

11. $35.7 \div 2.1$

12. $0.32 \div 0.4$

**Find each product.**

13. $6.65 \times 10{,}000$

14. $4.32 \times 0.01$

**Find each quotient.**

15. $2.374 \div 100$

16. $0.09 \div 0.01$

1. _____

2. _____

3. _____

4. _____

5. _____

6. _____

7. _____

8. _____

9. _____

10. _____

11. _____

12. _____

13. _____

14. _____

15. _____

16. _____

# Chapter Quiz

**Estimate each product.**

1. $49.2 \times 10.3$

2. $71.77 \times 6.1$

3. What is the value of $37.4 + 59.68$?

4. What is the value of $4.76 - 0.015$?

5. **SUBWAY** About how far could a subway car travel in 1 hour if its average speed was 0.96 mile per minute?

**Multiply.**

6. $1.2 \times 6$

7. $2.5 \times 4$

8. $3.1 \times 5$

9. $4.2 \times 6$

10. **GROCERIES** Oranges cost $0.25 each. How much will 7 oranges cost?

11. $3.02 \times 2.4$

12. $0.65 \times 0.25$

13. **GIFT WRAP** A rectangular piece of gift wrap is 5.5 inches by 2.1 inches. What is the area of the gift wrap? (*Hint*: Use $A = \ell w$.)

1. _____

2. _____

3. _____

4. _____

5. _____

6. _____

7. _____

8. _____

9. _____

10. _____

11. _____

12. _____

13. _____

# Vocabulary Test

| annex | dividend | product |
| compatible numbers | divisor | quotient |

## Fill in each blank with the correct term.

1. To estimate the quotient of decimals, use _____.

1. _____

2. In $25.83 \div 4.2$, 25.83 is the _____.

2. _____

3. The _____ of $35.6 \div 5$ is 7.12.

3. _____

4. The answer to a multiplication problem is called the _____.

4. _____

5. The answer to a division problem is called the _____.

5. _____

6. In $42 \div 2.1$, 2.1 is the _____.

6. _____

7. The _____ of $654.3 \times 0.001$ is 0.6543.

7. _____

8. The _____ of $59.4 \div 2$ is 59.4.

8. _____

## Define each term in your own words.

9. compatible numbers

9. _____

10. annex

10. _____

# Standardized Test Practice

**Read each question. Then fill in the correct answer on the answer sheet provided by your teacher or on a sheet or paper.**

1. Justin was estimating the area of the square sticky note shown below. Which would be a reasonable estimate of the area of the note? Use $A = lw$.

A. 36 square centimeters

B. 49 square centimeters

C. 64 square centimeters

D. 72 square centimeters

2. Marlene purchased 20 stamps at the post office for $8.40. What is the cost of one stamp?

   F. $0.41          H. $1.68

   G. $0.42          I. $2.38

3. [THINK SOLVE EXPLAIN] **SHORT RESPONSE** Manuel bought supplies for making party favors. The table shows the cost of each supply. If Manuel made 12 party favors, how much did it cost to make each party favor?

| Supply | Cost ($) |
|--------|----------|
| Bags | 2.00 |
| Candy | 5.75 |
| Stickers | 6.39 |
| Pencils | 4.82 |

4. ✏️ **GRIDDED RESPONSE** Shellie spent $10.56 on oranges that cost $0.88 per orange. How many oranges did she purchase?

5. ✏️ **GRIDDED RESPONSE** The average weight of a bass in a neighborhood lake is 5.1 pounds. Norman and his friend caught 4 bass. Assuming the bass were all average weight, what was the total weight in pounds of the fish they caught?

6. Bartholomew went hiking over the weekend. He hiked all 4 trails in 3 hours. Which is the **best** estimate for the number of miles he hiked per hour?

| Trail | Length (mi) |
|-------|-------------|
| Great Fork | 1.7 |
| Stoney Creek | 0.8 |
| Sippo Lake | 2.6 |
| Rock Falls | 0.5 |

   A. 2 miles          C. 15 miles

   B. 3 miles          D. 18 miles

7. The area of Trina's bedroom is 96.9 square feet. What is the length of her bedroom? Use $A = lw$.

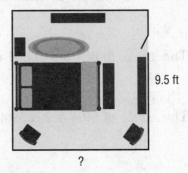

   F. 10.2 feet          H. 9.2 feet

   G. 10.0 feet          I. 9.0 feet

8. ✏️ **GRIDDED RESPONSE** Find the area in square meters of a rectangle with a length of 2.1 meters and a height of 0.8 meter.

9. A model plane is 100 times smaller than an actual plane. The length of the model is 4.8 inches. What is the actual length of the plane?

   A. 480 feet      C. 20 feet

   B. 40 feet      D. 8 feet

10. **THINK SOLVE EXPLAIN** **SHORT RESPONSE** Rita bought 5.7 pounds of bananas and 2.8 pounds of apples. Write a multiplication expression and find the total cost for the fruit. Round to the nearest cent.

Bananas
$ 0.59 per pound
Apples
$ 1.99 per pound

11. Malabar Middle School is raising money for a local charity. Their goal is to raise $500 by their holiday break. If they have 10 days before their break, what is a reasonable amount that they should collect each day to reach their goal?

   F. $5      H. $50

   G. $25      I. $100

12. Ignacio cut the board shown into 4.5-inch pieces. How many pieces can he cut?

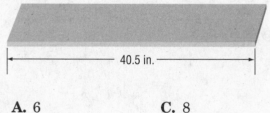

   ←————— 40.5 in. —————→

   A. 6      C. 8
   B. 7      D. 9

13. The table below shows times from the Men's $4 \times 100$ meter Medley. The four swimmers each swam 100 meters.

| 2008 Olympic Results | |
|---|---|
| **Team** | **Time (min:sec)** |
| United States | 3:29.34 |
| Australia | 3:30.04 |
| Japan | 3:31.18 |
| Russian Federation | 3:31.92 |

Suppose each swimmer swam the same amount of time. How long did each United States swimmer swim in the race?

   F. 67.335 s      H. 52.335 s

   G. 67 s      I. 52 s

14. **THINK SOLVE EXPLAIN** **EXTENDED RESPONSE** Wesley lives in an apartment and would like to have his own vegetable garden. His city offers garden plots, but he has to pay for the fencing. Use $A = lw$.

| Plot A | Plot B | Plot C |
|---|---|---|
| Area: 204 ft² | Area: ▪ | Area: 204.49 ft² |
| Length: ▪ | Length: 12.78 ft | Length: ▪ |
| Height: 10.2 ft | Height: ▪ | Height: 14.3 ft |
| Perimeter: ▪ | Perimeter: 57.56 ft | Perimeter: ▪ |

**Part A** Find the missing measurement for each plot.

**Part B** Which plot requires the least amount of fencing?

**Part C** Which plot has the greatest area?

# Student Recording Sheet

*Use this recording sheet with the Standardized Test Practice pages.*

**Fill in the correct answer. For gridded-response questions, write your answers in the boxes on the answer grid and fill in the bubbles to match your answers.**

1. Ⓐ Ⓑ Ⓒ Ⓓ

2. Ⓕ Ⓖ Ⓗ Ⓘ

3. _____

4.

5.

6. Ⓐ Ⓑ Ⓒ Ⓓ

7. Ⓕ Ⓖ Ⓗ Ⓘ

8.

9. Ⓐ Ⓑ Ⓒ Ⓓ

10. _____

11. Ⓕ Ⓖ Ⓗ Ⓘ

12. Ⓐ Ⓑ Ⓒ Ⓓ

13. Ⓕ Ⓖ Ⓗ Ⓘ

## Extended Response

Record your answers for Exercise 14 on the back of this paper.

# Extended-Response Test

**Demonstrate your knowledge by giving a clear, concise solution to each problem. Be sure to include all relevant drawings and justify your answers. You may show your solutions in more than one way or investigate beyond the requirements of the problem. If necessary, record your answers on another piece of paper.**

1. A surveyor is surveying a conservation area.

   **a.** A rectangular wetland conservation area is 5.25 miles by 4.8 miles. Estimate the area of the park. Use $A = \ell w$. Explain each step.

   **b.** Explain how to multiply decimals. Use the dimensions in part **a** to find the area of the conservation area.

   **c.** Explain how to divide decimals. Then divide to find the length of a rectangular lot that is 1.6 miles wide and has an area of 2.8 square miles.

2. The people of a neighborhood are creating a community vegetable garden.

   **a.** One part of the garden will have 6 rows of corn plants. Each row will be 11.25 feet long. Irrigation soaker hoses will be laid along each row of corn plants. How many feet of soaker hose will be needed? Explain the steps you go through to solve this problem. Be sure to explain how you decide where to place the decimal point.

   **b.** The neighbors have $18.57 to spend on strawberry plants. A 6-pack of plants costs $2. How many packs can they buy? Show your work.

3. **a.** Explain how to complete the equation
   $3.694 \times$ _____ $= 36,940$.

   **b.** Explain why it might be necessary to annex zeros when dividing decimals.

# Extended-Response Rubric

| Score | Description |
|:---:|:---|
| 4 | A score of four is a response in which the student demonstrates a thorough understanding of the mathematics concepts and/or procedures embodied in the task. The student has responded correctly to the task, used mathematically sound procedures, and provided clear and complete explanations and interpretations. The response may contain minor flaws that do not detract from the demonstration of a thorough understanding. |
| 3 | A score of three is a response in which the student demonstrates an understanding of the mathematics concepts and/or procedures embodied in the task. The student's response to the task is essentially correct with the mathematical procedures used and the explanations and interpretations provided demonstrating an essential but less than thorough understanding. The response may contain minor flaws that reflect inattentive execution of mathematical procedures or indications of some misunderstanding of the underlying mathematics concepts and/or procedures. |
| 2 | A score of two indicates that the student has demonstrated only a partial understanding of the mathematics concepts and/or procedures embodied in the task. Although the student may have used the correct approach to obtaining a solution or may have provided a correct solution, the student's work lacks an essential understanding of the underlying mathematical concepts. The response contains errors related to misunderstanding important aspects of the task, misuse of mathematical procedures, or faulty interpretations of results. |
| 1 | A score of one indicates that the student has demonstrated a very limited understanding of the mathematics concepts and/or procedures embodied in the task. The student's response is incomplete and exhibits many flaws. Although the student's response has addressed some of the conditions of the task, the student reached an inadequate conclusion and/or provided reasoning that was faulty or incomplete. The response exhibits many flaws or may be incomplete. |
| 0 | A score of zero indicates that the student has provided no response at all, or a completely incorrect or uninterpretable response, or demonstrated insufficient understanding of the mathematics concepts and/or procedures embodied in the task. For example, a student may provide some work that is mathematically correct, but the work does not demonstrate even a rudimentary understanding of the primary focus of the task. |

# Test, Form 1A

**Write the letter for the correct answer in the blank at the right of each question.**

**For Exercises 1–2, which is the best estimate for each product?**

1. $6.8 \times 2.3$

    A. 7          B. 12          C. 14          D. 140          1. _____

2. $25.4 \times 5.1$

    F. 100          G. 125          H. 150          I. 175          2. _____

3. Alicia bought 3.8 pounds of cashews for $15.74. About how much was the cost per pound?

    A. $3          B. $4          C. $5          D. $6          3. _____

4. Mr. Stanton bought 2.6 yards of fabric for $8.94. About how much was the cost per yard?

    F. $1          G. $2          H. $3          I. $4          4. _____

5. A hummingbird egg's mass is approximately 0.374 milligram. What is the mass of 4 hummingbird eggs?

    A. 0.1496 mg     B. 1.496 mg     C. 14.96 mg     D. 149.6 mg     5. _____

6. What is the value of $6 \times 333$?

    F. 1,888          G. 1,898          H. 1,980          I. 1,998          6. _____

7. What is the value of $2.7 \times 4.1$?

    A. 1.107          B. 11.07          C. 110.7          D. 1,107          7. _____

8. What is the value of $5.87 \times 1.5$?

    F. 8.805          G. 88.05          H. 880.5          I. 8,805          8. _____

9. The dance committee has 1,036 flowers. They want to divide the flowers evenly among 28 centerpieces. How many flowers will be in each centerpiece?

    A. 28 flowers     B. 34 flowers     C. 37 flowers     D. 40 flowers     9. _____

10. The student council sold 362 roles of wrapping paper for a fundraiser. If they raised $4,706, what was the cost of one role of wrapping paper?

    F. $9          G. $11          H. $13          I. $15          10. _____

11. What is the next term in the pattern shown below?
    2.1, 3.2, 4.3, 5.4, ...

    A. 1.1          C. 5.5

    B. 4.3          D. 6.5          11. _____

# Test, Form 1A    *(continued)*

**12.** What is the value of 145.3 × 0.1?

    **F.** 0.1453    **G.** 1.453    **H.** 14.53    **I.** 145.3    **12.** _____

**Find each quotient.**

**13.** 12.6 ÷ 9

    **A.** 0.0014    **B.** 0.014    **C.** 0.14    **D.** 1.4    **13.** _____

**14.** $23\overline{)1863}$

    **F.** 73    **G.** 81    **H.** 87    **I.** 90    **14.** _____

**15.** One meter is approximately equal to 39.39 inches. About how many inches are in 3 meters?

    **A.** 40    **B.** 90    **C.** 120    **D.** 150    **15.** _____

**16.** What is the value of the expression 54.36 ÷ 0.1?

    **F.** 543.6    **G.** 54.36    **H.** 5.436    **I.** 0.5436    **16.** _____

**17.** The sixth-grade class is going to the local art museum. There are 46 students in the class. The total cost of the tickets for all the students is $379.50. What is the cost of one student ticket?

    **A.** $8    **B.** $8.25    **C.** $8.50    **D.** $9    **17.** _____

**18.** Alison bought one T-shirt for $10.20 and another T-shirt for $18.90. What is the total amount she spent?

    **F.** $8.70    **G.** $28.10    **H.** $29.10    **I.** $30.00    **18.** _____

**19.** A city had 5.4 inches of rain in September and 4.5 inches of rain in October. What is the difference in rainfall between the two months?

    **A.** 9.9        **C.** 0.9

    **B.** 1         **D.** 0    **19.** _____

**20.** Tristan had 95 inches of ribbon. He cut off 27.50 inches. How many inches of ribbon remain?

    **F.** 122.50 in.        **H.** 70 in.

    **G.** 100.50 in.        **I.** 67.50 in.    **20.** _____

# Test, Form 1B

**Write the letter for the correct answer in the blank at the right of each question.**

**For Exercises 1–2, which is the best estimate for each product?**

1. $9.2 \times 6.8$
   **A.** 60      **B.** 63      **C.** 68      **D.** 630

   1. _____

2. $11.2 \times 1.73$
   **F.** 15      **G.** 22      **H.** 24      **I.** 34

   2. _____

3. Kevin bought 4.6 pounds of pecans for $50.89. About how much was the cost per pound?
   **A.** $5      **B.** $8      **C.** $10      **D.** $20

   3. _____

4. Mrs. Garcia bought 9.1 yards of fabric for $18.79 . About how much was the cost per yard?
   **F.** $1      **G.** $2      **H.** $3      **I.** $4

   4. _____

5. A sea horse travels an average of 0.01 mile per hour. At that rate, how far can a sea horse travel in 2.4 hours?
   **A.** 0.024 mi      **B.** 2.4 mi      **C.** 24 mi      **D.** 240 mi

   5. _____

6. What is the value of $5 \times 108$?
   **F.** 113      **G.** 432      **H.** 535      **I.** 540

   6. _____

7. What is the value of $3.6 \times 4.2$?
   **A.** 0.01512      **B.** 0.1512      **C.** 1.512      **D.** 15.12

   7. _____

8. What is the value of $7.24 \times 0.6$?
   **F.** 0.4344      **G.** 4.344      **H.** 43.44      **I.** 434.4

   8. _____

9. The dance committee has 1,120 flowers. They want to divide the flowers evenly among 32 centerpieces. How many flowers will be in each centerpiece?
   **A.** 20 flowers      **B.** 25 flowers      **C.** 30 flowers      **D.** 35 flowers

   9. _____

10. The student council sold 286 roles of wrapping paper for a fundraiser. If they raised $3,146, what was the cost of one role of wrapping paper?
    **F.** $9      **G.** $11      **H.** $13      **I.** $14

    10. _____

11. What is the next term in the pattern shown below?
    10, 8.9, 7.8, 6.7, ...
    **A.** 6.6      **C.** 5.5
    **B.** 5.6      **D.** 1.1

    11. _____

# Test, Form 1B   (continued)

**12.** What is the value of 56.99 × 0.01?

    **F.** 569.9      **G.** 56.99      **H.** 5.699      **I.** 0.5699      **12.** _____

**Find each quotient.**

**13.** 12.8 ÷ 4

    **A.** 32      **B.** 3.2      **C.** 0.32      **D.** 0.032      **13.** _____

**14.** $23\overline{)667}$

    **F.** 32      **G.** 31      **H.** 29      **I.** 28      **14.** _____

**15.** One meter is approximately equal to 39.39 inches. About how many inches are in 5 meters?

    **A.** 80      **B.** 120      **C.** 150      **D.** 200      **15.** _____

**16.** What is the value of the expression 75.42 ÷ 0.1?

    **F.** 754.2      **G.** 75.42      **H.** 7.542      **I.** 0.7542      **16.** _____

**17.** The sixth-grade class is going to the local art museum's dinosaur exhibit. There are 52 students in the class. The total cost of the tickets for all the students is $364. What is the cost of one student ticket?

    **A.** $6      **B.** $7      **C.** $7.50      **D.** $8      **17.** _____

**18.** Wesley bought one T-shirt for $7.90 and another T-shirt for $12.40. What is the total amount he spent?

    **F.** $20.30      **G.** $20      **H.** $4.50      **I.** $4      **18.** _____

**19.** A city had 9.2 inches of rain in April and 4.3 inches of rain in May. What is the difference in rainfall between the two months?

    **A.** 5                **C.** 3.9

    **B.** 4.9            **D.** 3      **19.** _____

**20.** Juanna had a board that was 54 inches long. She cut off 22.5 inches for a project. How many inches of the board remain?

    **F.** 30 in.      **G.** 31.5 in.      **H.** 32.5 in.      **I.** 40 in.      **20.** _____

# Test, Form 2A

**Write the letter for the correct answer in the blank at the right of each question.**

**For Exercises 1–4, what is the best estimate for each product or quotient?**

1. $8.3 \times 4.9$

   **A.** 40    **B.** 45    **C.** 48    **D.** 54

   1. _____

2. $12.1 \times 2.95$

   **F.** 22    **G.** 24    **H.** 36    **I.** 39

   2. _____

3. $60.94 \div 5.7$

   **A.** 5    **B.** 8    **C.** 10    **D.** 20

   3. _____

4. $32.79 \div 8.1$

   **F.** 1    **G.** 2    **H.** 3    **I.** 4

   4. _____

5. A cat can jump five times as high as it is tall. How high could a cat jump if it is 16.4 centimeters tall?

   **A.** 8.2 cm    **B.** 82 cm    **C.** 820 cm    **D.** 8,200 cm

   5. _____

6. The temperature in a city on a Monday was 78.3 degrees Fahrenheit. On Tuesday, the temperature was 87.1 degrees Fahrenheit. How much higher was the temperature on Tuesday?

   **F.** 8 degrees          **H.** 9.2 degrees

   **G.** 8.8 degrees        **I.** 9.8 degrees

   6. _____

7. Bananas cost $0.54 per pound and grapes cost $1.28 per pound. Leanne bought 2.6 pounds of bananas and 3.1 pounds of grapes. How much did she pay for the bananas and grapes?

   **A.** $1.82    **B.** $2.68    **C.** $4.54    **D.** $5.37

   7. _____

8. A family travels 989.5 miles in 17.8 hours. Estimate the number of miles they can travel in one hour.

   **F.** 25 miles    **G.** 40 miles    **H.** 50 miles    **I.** 65 miles

   8. _____

9. Callie has 1,012 beads. She wants to divide the beads evenly to make 22 necklaces. How many beads will be used for each necklace?

   **A.** 22 beads    **B.** 42 beads    **C.** 46 beads    **D.** 50 beads

   9. _____

10. On Saturday, Hattie biked 3.2 hours at a speed of 12.5 miles per hour. On Sunday, she biked 2.5 hours at a speed of 14.3 miles per hour. How much farther did she bike on Saturday?

    **F.** 3 miles           **H.** 5.75 miles

    **G.** 4.25 miles        **I.** 14.51 miles

    10. _____

# Test, Form 2A    *(continued)*

**11.** What is the next term in the pattern shown below?

4.5, 9, 13.5, 18, ...

**A.** 4.5                          **C.** 22.5

**B.** 22.4                        **D.** 23

11. _____

**Divide.**

**12.** $4.5\overline{)38.7}$

**F.** 0.86          **G.** 8.6          **H.** 86          **I.** 860

12. _____

**13.** $604 \div 4$

**A.** 151          **B.** 160          **C.** 302          **D.** 600

13. _____

**14.** Ling bought 22 boxes of crayons for $28.16. How much did she pay for each box of crayons?

14. _____

**15.** Evaluate $252.7 \div 19$.

15. _____

**16.** Determine the missing factor in ■ $\times 4.2 = 16.8$.

16. _____

**Find the area of each figure.** (*Hint*: Use $A = lw$.)

**17.**

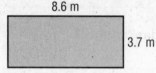

8.6 m

3.7 m

**18.**

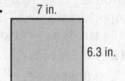

7 in.

6.3 in.

17. _____

18. _____

**19.** Isidro wants to purchase a used scooter that costs $79.95. The sales tax is found by multiplying the price of the scooter by 0.065. Find the total amount of money Isidro will pay. Round to the nearest cent.

19. _____

**20.** There are 632 students going on a school field trip. Each bus can carry 48 students. How many buses should the school reserve? Explain your reasoning.

20. _____

**21.** Mr. Shaw wants to replace the flooring his family room. The floor has an area of 262.8 square feet. If the room is 18 feet long, how wide is it? Justify your procedure.

21. _____

**22.** Nina's sister has 386 stickers that she wants to put in a book. Each page can hold 42 stickers. How many pages will she need? Explain your reasoning.

22. _____

# Test, Form 2B

**Write the letter for the correct answer in the blank at the right of each question.**

**For Exercises 1–4, what is the best estimate for each product or quotient?**

1. $6.3 \times 7.7$

   **A.** 42      **B.** 48      **C.** 54      **D.** 56      1. _____

2. $11.2 \times 3.95$

   **F.** 33      **G.** 36      **H.** 44      **I.** 48      2. _____

3. $71.94 \div 5.7$

   **A.** 10      **B.** 12      **C.** 15      **D.** 18      3. _____

4. $35.69 \div 7.1$

   **F.** 3      **G.** 4      **H.** 5      **I.** 6      4. _____

5. A cat can jump five times as high as it is tall. How high could a cat jump if it were 17.4 centimeters tall?

   **A.** 8.7 cm      **B.** 87 cm      **C.** 870 cm      **D.** 8,700 cm      5. _____

6. The temperature in a city on a Friday was 46.8 degrees Fahrenheit. On Saturday, the temperature was 63.5 degrees Fahrenheit. How much higher was the temperature on Saturday?

   **F.** 15 degrees      **H.** 17.3 degrees

   **G.** 16.7 degrees      **I.** 18.5 degrees      6. _____

7. Bananas cost $0.49 per pound and grapes cost $1.15 per pound. Leanne bought 3.5 pounds of bananas and 2.4 pounds of grapes. How much did she pay for the bananas and grapes?

   **A.** $2.76      **B.** $4      **C.** $4.48      **D.** $5.52      7. _____

8. A family travels 613.4 miles in 11.5 hours. Estimate the number of miles they can travel in one hour.

   **F.** 40 miles      **G.** 50 miles      **H.** 65 miles      **I.** 70 miles      8. _____

9. Ryan has 782 beads. He wants to divide the beads evenly to make 34 bracelets. How many beads will be used for each bracelet?

   **A.** 20 beads      **B.** 23 beads      **C.** 26 beads      **D.** 34 beads      9. _____

10. On Saturday, Mason biked 1.5 hours at a speed of 11.9 miles per hour. On Sunday, he biked 3.2 hours at a speed of 14.8 miles per hour. How much farther did he bike on Sunday?

    **F.** 2.9 miles      **H.** 26.7 miles

    **G.** 17.85 miles      **I.** 29.51 miles      10. _____

# Test, Form 2B    (continued)

11. What is the next term in the pattern shown below?

     12.5, 25, 37.5, 50, ...

     **A.** 75                     **C.** 13

     **B.** 62.5                   **D.** 12.50                       11. _____

**Divide.**

12. 4.5)‾39.6‾

     **F.** 880      **G.** 88.0      **H.** 8.8      **I.** 0.88     12. _____

13. 806 ÷ 4

     **A.** 100.75   **B.** 200      **C.** 201.5    **D.** 403       13. _____

14. David bought 26 boxes of crayons for $38.48. How much did he pay
    for each box of crayons?                                         14. _____

15. Evaluate 292.4 ÷ 17.                                             15. _____

16. Determine the missing factor in ■ × 4.3 = 12.9.                  16. _____

**Find the area of each figure. (*Hint*: Use $A = lw$.)**

17.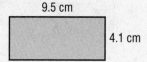
     9.5 cm

     4.1 cm                                                          17. _____

18.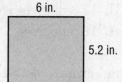
     6 in.

     5.2 in.                                                         18. _____

19. Isidro wants to purchase a used scooter that costs $89.75. The sales
    tax is found by multiplying the price of the scooter by 0.065. Find
    the total amount of money Isidro will pay. Round to the nearest cent.  19. _____

20. There are 745 students going on a school field trip. Each bus can
    carry 46 students. How many buses should the school reserve?
    Explain your reasoning.                                          20. _____

21. Mr. Thomas wants to replace the flooring his den. The floor has an
    area of 201.6 square feet. If the room is 16 feet long, how wide is it?
    Justify your procedure.                                          21. _____

22. Molly's sister has 475 stickers that she wants to put in a book. Each
    page can hold 38 stickers. How many pages will she need? Explain
    your reasoning.                                                  22. _____

# Test, Form 3A

**For Exercises 1–6, estimate each product or quotient.**

1. $6.2 \times 9.1$

2. $12.4 \times 2.8$

3. $25.2 \div 7.8$

4. $9.3\overline{)62.1}$

5. A car can travel 24.8 miles for every 1 gallon of gasoline. Estimate the number of miles the car can travel on 9.5 gallons.

6. A roll of wire has 43 inches of wire. If each bracelet requires 6.8 inches of wire, estimate the number of bracelets that can be made. Explain why your estimate is reasonable.

7. Jackson ran 3.6 miles a day. How many miles did he run in 4 days?

8. A boxed lunch costs $7.01. How much would 9 boxed lunches cost?

9. A flower bed is 0.5 meter wide and 2.7 meters long. What is the area? (*Hint*: Use $A = lw$.)

10. A ball rolls 5.24 feet each second. How far will it roll in 1.8 seconds?

11. The prices for recycling aluminum and copper are shown in the table. What is the total price of 2.5 pounds of aluminum and 3 pounds of copper?

| Item | Price ($/lb) |
| --- | --- |
| aluminum | 0.46 |
| copper | 1.08 |

1. _____

2. _____

3. _____

4. _____

5. _____

6. _____

7. _____

8. _____

9. _____

10. _____

11. _____

**Find each product or quotient.**

12. $0.21 \times 10$

13. $45.35 \div 10,000$

14. $6.02 \times 0.1$

15. $9.4 \div 0.01$

16. What is the next term in the pattern 3.1, 6.2, 9.3, 12.4, ...?

17. A disposable camera costs $6.95. A party planner wants to buy 100 cameras for an upcoming event. How much will they cost?

12. _____

13. _____

14. _____

15. _____

16. _____

17. _____

# Test, Form 3A  (continued)

**18.** Alisha earned $5.50 mowing the lawn and $5.50 raking the leaves. How much total money did she earn?

18. _____

**19.** Jacob had $4.30 and spent $2.90 on snacks. How much did he have left?

19. _____

**Divide.**

**20.** $0.2\overline{)6.2}$

20. _____

**21.** $38.07 \div 9.4$

21. _____

**22.** $50.43 \div 6.15$

22. _____

**23.** $77.5 \div 2.5$

23. _____

**24.** $33.33 \div 6.6$

24. _____

**25.** The cost of an online movie rental service is $89.94 for 6 months. What is the cost per month?

25. _____

**26.** If 4 people are going to share 14.8 ounces of cheese, how much will each person get?

26. _____

**27.** The three-toed sloth can travel faster in the trees than on the ground, as shown in the table. If a sloth traveled for 2.5 hours, how much farther could it go in the trees than on the ground?

27. _____

| Location | Average Speed (mi/h) |
|----------|----------------------|
| On ground | 0.08 |
| In trees | 0.17 |

**28.** A new workout facility needs 925 rubber floor tiles. If the total cost for the rubber floor tiles is $5,550, how much does one floor tile cost?

28. _____

**29.** A large cake from a bakery will serve 60 people. If there are 832 people expected at a reception, how many cakes are needed? Explain your reasoning.

29. _____

**30.** Movie tickets cost $9.95 for adults and $6.75 for students. The Ryder family buys 2 adult tickets, 3 student tickets, and then popcorn and drinks for $26. What is the total amount that they spend?

30. _____

# Test, Form 3B

**For Exercises 1–4, estimate each product or quotient.**

1. $5.9 \times 3.1$

1. _____

2. $6.1 \times 10.9$

2. _____

3. $35.1 \div 7.8$

3. _____

4. $8.2\overline{)56.1}$

4. _____

5. A car can travel 19.6 miles for every 1 gallon of gasoline. Estimate the number of miles the car can travel on 13.8 gallons.

5. _____

6. A roll of wire has 51 inches of wire. If each bracelet requires 7.2 inches of wire, estimate the number of bracelets that can be made. Explain why your estimate is reasonable.

6. _____

7. Olivia ran 2.4 miles a day. How many miles did she run in 4 days?

7. _____

8. A boxed lunch costs $8.49. How much would 8 boxed lunches cost?

8. _____

9. A flower bed is 0.8 meter wide and 3.4 meters long. What is the area? (*Hint*: Use $A = lw$.)

9. _____

10. A ball rolls 5.15 feet each second. How far will it roll in 2.6 seconds?

10. _____

11. The prices for recycling aluminum and copper are shown in the table. What is the total price of 3 pounds of aluminum and 2.5 pounds of copper?

| Item | Price ($/lb) |
|---|---|
| aluminum | 0.46 |
| copper | 1.09 |

11. _____

**Find each product or quotient.**

12. $0.68 \times 100$

12. _____

13. $24.92 \div 1,000$

13. _____

14. $0.4 \times 0.1$

14. _____

15. $7.5 \div 0.01$

15. _____

16. What is the next term in the pattern, 4.4, 8.8, 13.2, 17.6, ...?

16. _____

17. A disposable camera costs $5.95. A party planner wants to buy 1,000 cameras for upcoming events. How much will they cost?

17. _____

# Test, Form 3B   (continued)

**18.** Denzel earned $8.50 mowing the lawn and $6 raking the leaves. How much total money did he earn?

18. _____

**19.** Jaquie had $6.58 and spent $3.25 on snacks. How much did she have left?

19. _____

**Divide.**

**20.** $0.3\overline{)7.2}$

20. _____

**21.** $59.84 \div 8.5$

21. _____

**22.** $48.62 \div 7.15$

22. _____

**23.** $3.399 \div 6.6$

23. _____

**24.** $82.5 \div 2.5$

24. _____

**25.** The cost of a gym membership is $173.94 for 6 months. What is the cost per month?

25. _____

**26.** If 5 people are going to share 16.75 ounces of cheese, how much will each person get?

26. _____

**27.** The three-toed sloth can travel faster in the trees than on the ground, as shown in the table. If a sloth traveled for 3.5 hours, how much farther could it go in the trees than on the ground?

| Location | Average Speed (mi/h) |
|----------|----------------------|
| On ground | 0.08 |
| In trees | 0.17 |

27. _____

**28.** A new workout facility needs 1,020 rubber floor tiles. If the total cost for the rubber floor tiles is $8,160, how much does one floor tile cost?

28. _____

**29.** A large cake from a bakery will serve 55 people. If there are 725 people expected at a reception, how many cakes are needed? Explain your reasoning.

29. _____

**30.** Movie tickets cost $9.95 for adults and $6.75 for students. The Ryder family buys 2 adult tickets, 4 student tickets, and then popcorn and drinks for $28. What is the total amount that they spend?

30. _____

# Are You Ready?

## Review

To estimate sums and differences of mixed numbers, round each mixed number to the nearest whole number.

### Example 1

Estimate $3\frac{1}{6} + 2\frac{5}{8}$.

$\frac{1}{6} < \frac{1}{2}$, so $3\frac{1}{6}$ rounds down to 3.

$\frac{5}{8} > \frac{1}{2}$, so $2\frac{5}{8}$ rounds up to 3.

$$3 + 3 = 6$$

So, $3\frac{1}{6} + 2\frac{5}{8}$ is *about* 6.

### Example 2

Estimate $8\frac{5}{6} - 3\frac{2}{3}$.

$\frac{5}{6} > \frac{1}{2}$, so $8\frac{5}{6}$ rounds up to 9.

$\frac{2}{3} > \frac{1}{2}$, so $3\frac{2}{3}$ rounds up to 4.

$$9 - 4 = 5$$

So, $8\frac{5}{6} - 3\frac{2}{3}$ is *about* 5.

## Exercises

**Estimate.**

1. $8\frac{1}{5} + 9\frac{5}{8}$

2. $5\frac{7}{12} + 1\frac{3}{8}$

3. $4\frac{1}{6} + 11\frac{3}{10}$

4. $5\frac{1}{6} + 4\frac{1}{3}$

5. $10\frac{1}{5} - 9\frac{4}{7}$

6. $9\frac{9}{15} - 3\frac{7}{8}$

7. $7\frac{2}{3} - 2\frac{6}{7}$

8. $6\frac{7}{8} - 3\frac{1}{4}$

1. _____

2. _____

3. _____

4. _____

5. _____

6. _____

7. _____

8. _____

# Are You Ready?

## Practice

**Estimate.**

1. $7\frac{7}{8} + 5\frac{1}{5}$

2. $6\frac{5}{6} + 4\frac{5}{8}$

3. $3\frac{3}{4} + 15\frac{1}{9}$

4. $11\frac{1}{7} - 9\frac{1}{8}$

5. $12\frac{3}{5} - 6\frac{2}{3}$

6. **HEIGHTS** Johanna is $63\frac{1}{4}$ inches tall and Oleta is $66\frac{3}{4}$ inches tall. About how much taller is Oleta than Johanna?

**Add or subtract. Write in simplest form.**

7. $\frac{1}{2} + \frac{5}{6}$

8. $\frac{2}{3} + \frac{7}{9}$

9. $\frac{7}{16} + \frac{3}{4}$

10. $6\frac{3}{4} - 2\frac{1}{2}$

11. $10\frac{7}{10} - 5\frac{3}{5}$

12. $3\frac{4}{5} + 2\frac{1}{3}$

13. $3\frac{1}{2} + 2\frac{1}{3}$

14. **BIKING** Kiyo biked $5\frac{3}{8}$ miles on Saturday and $3\frac{3}{4}$ miles on Sunday. How far did she bike in all?

1. _____

2. _____

3. _____

4. _____

5. _____

6. _____

7. _____

8. _____

9. _____

10. _____

11. _____

12. _____

13. _____

14. _____

# Are You Ready?

## Apply

---

**1. MUSIC** Kristy downloads two songs to her MP3 player. The songs are $3\frac{3}{10}$ minutes and $4\frac{2}{3}$ minutes long. About how many minutes of memory will these two songs use altogether?

**2. TRAINING** Mike ran $14\frac{1}{5}$ miles during his first week of training and $25\frac{3}{8}$ miles during his second week. About how many miles has he run in all?

---

**3. CATS** In a cat shelter, the longest cat is $18\frac{7}{8}$ inches long and the shortest cat is $13\frac{1}{4}$ inches long. About how much longer is the longest cat than the shortest cat?

**4. WALKING** Monette's piano instructor's house is $7\frac{1}{8}$ miles from her house. After biking for $1\frac{3}{4}$ miles, she stops to rest. How much longer does she need to bike to reach her instructor's house?

---

**5. READING** Carole read $\frac{1}{8}$ of her book on Monday and $\frac{1}{6}$ of her book on Tuesday. What fraction of her book did she read altogether on these days?

**6. ROPE** Mr. Silva cut $3\frac{9}{16}$ feet of rope from a bundle that was $10\frac{3}{4}$ feet long. What length of rope remains in the bundle?

---

# Diagnostic Test

**Estimate.**

1. $5\frac{7}{8} + 4\frac{1}{5}$

2. $7\frac{1}{6} + 4\frac{7}{8}$

3. $3\frac{3}{4} + 6\frac{4}{5}$

4. $10\frac{1}{8} - 9\frac{1}{9}$

1. _____

2. _____

3. _____

4. _____

5. **GASOLINE** While on a jet ski, Maddy used $2\frac{3}{4}$ gallons of gasoline on Tuesday and $2\frac{1}{5}$ gallons of gasoline on Wednesday. About how many gallons of gas did she use altogether?

5. _____

**Add or subtract. Write in simplest form.**

6. $\frac{8}{9} - \frac{1}{3}$

7. $\frac{3}{4} + \frac{1}{8}$

8. $\frac{6}{7} - \frac{3}{4}$

9. $5\frac{9}{10} - 3\frac{1}{5}$

10. $7\frac{4}{9} + 3\frac{2}{3}$

11. $8\frac{5}{6} - 3\frac{1}{3}$

12. $10\frac{2}{3} + 8\frac{7}{10}$

6. _____

7. _____

8. _____

9. _____

10. _____

11. _____

12. _____

13. **FLOUR** A chef has $3\frac{1}{16}$ pounds of flour in his kitchen. He buys another $6\frac{1}{2}$ pounds of flour. How much flour does he have in all?

13. _____

# Pretest

### Estimate each product.

1. $\frac{1}{5} \times 24$

2. $\frac{7}{8} \times \frac{3}{5}$

3. $3\frac{7}{8} \times 10\frac{1}{10}$

1. _____

2. _____

3. _____

### Multiply. Write in simplest form.

4. $6 \times \frac{2}{3}$

5. $\frac{4}{5} \times 15$

6. $\frac{1}{4} \times \frac{5}{6}$

7. $\frac{1}{3} \times 1\frac{1}{3}$

4. _____

5. _____

6. _____

7. _____

### Find the reciprocal of each number.

8. $\frac{1}{2}$

9. $4$

10. $\frac{3}{5}$

8. _____

9. _____

10. _____

### Divide. Write in simplest form.

11. $3 \div \frac{2}{5}$

12. $\frac{4}{5} \div \frac{1}{2}$

13. $\frac{9}{10} \div 9$

14. $9 \div 1\frac{1}{9}$

15. $2\frac{1}{2} \div 3\frac{1}{3}$

16. $4\frac{1}{2} \div 2\frac{7}{10}$

11. _____

12. _____

13. _____

14. _____

15. _____

16. _____

# Chapter Quiz

**Estimate each product.**

1. $\frac{1}{3} \times 28$

1. _____

2. $\frac{4}{5} \times 1\frac{8}{9}$

2. _____

3. $4\frac{2}{5} \times 2\frac{2}{3}$

3. _____

4. $2\frac{6}{7} \times 3\frac{1}{4}$

4. _____

5. A reservoir is $3\frac{1}{4}$ miles wide. Tony sailed $\frac{4}{5}$ the width of the reservoir. About how far did Tony sail?

5. _____

**Multiply. Write in simplest form.**

6. $6 \times \frac{2}{3}$

6. _____

7. $\frac{2}{7} \times 9$

7. _____

8. $\frac{5}{6} \times 12$

8. _____

9. **MUFFINS** There are 24 muffins in a bakery. If $\frac{3}{8}$ of the muffins are blueberry, how many muffins are blueberry?

9. _____

10. How many feet are in 27 yards?

10. _____

11. Kenya picked 18 flowers from her garden. If $\frac{3}{5}$ of flowers in her garden remain unpicked, how many flowers are still in her garden?

11. _____

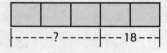

# Vocabulary Test

| | |
|---|---|
| Commutative Property | mixed number |
| denominator | numerator |
| dimensional analysis | reciprocal |
| fraction | simplest form |
| improper fraction | unit ratio |

## Choose from the terms above to complete each sentence.

1. A number that represents part of a whole or part of a set is called a(n) _____.

    1. _____

2. When dividing fractions, it is most helpful to change a mixed number into a(n) _____.

    2. _____

3. The _____ of a fraction is the part that tells how many units the fraction contains.

    3. _____

4. The _____ states that the order in which numbers are multiplied does not change the product.

    4. _____

5. A(n) _____ is made up of a whole number and a fraction.

    5. _____

6. In $\frac{4}{5}$, 5 is the _____.

    6. _____

7. In a _____, the denominator is 1 unit.

    7. _____

8. The _____ of 2 is $\frac{1}{2}$.

    8. _____

## Define each term in your own words.

9. dimensional analysis

    9. _____

10. simplest form

    10. _____

# Standardized Test Practice

**Read each question. Then fill in the correct answer on the answer sheet provided by your teacher or on a sheet of paper.**

1. An indoor soccer goal is $6\frac{1}{2}$ feet high. An outdoor soccer goal is $1\frac{3}{13}$ times as high as an indoor goal. How high is an outdoor goal?

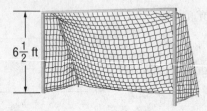

$6\frac{1}{2}$ ft

   **A.** 6 feet      **C.** 8 feet
   **B.** 7 feet      **D.** 9 feet

2. Drew read 120 pages of his book, which was 0.60 of the book. How many pages are in the book?

   **F.** 7.2      **H.** 72
   **G.** 20      **I.** 200

3. How many times longer is the length of the rectangle compared to the width?

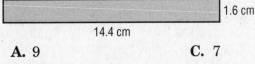

1.6 cm
14.4 cm

   **A.** 9      **C.** 7
   **B.** 8      **D.** 6

4. ✎ **GRIDDED RESPONSE** Estimate the cost in dollars of 4 boxes of cereal.

Puffs Cereal
$2.89

5. ✎ **GRIDDED RESPONSE** Reggie spent $\frac{2}{5}$ of his day at a day camp exploring nature. If he was at camp for 10 hours, how many hours did he spend exploring nature?

6. Margaret has saved $400. She is going to spend one tenth of her money on a shopping spree. How much money will Margaret spend?

   **F.** $0.40
   **G.** $4
   **H.** $40
   **I.** $100

7. Lacie wants to paint her bedroom walls. She has to cover 1,162.5 square feet. About how many gallons of paint will she need?

| Gallons | Coverage (ft²) |
|---------|----------------|
| 1 | 250 |
| 2 | 500 |
| 3 | 750 |

   **A.** 5
   **B.** 4
   **C.** 3
   **D.** 2

8. ✎ **GRIDDED RESPONSE** The Brighams' yard is shown. If they divide the yard into 3 equal sections, what is the size, in acres, of each section?

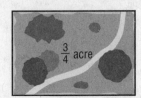

$\frac{3}{4}$ acre

9. Miguela bought oranges and grapes. Use the table to find the cost of 3.4 pounds of oranges and 2.8 pounds of grapes.

| Fruit | Price per Pound ($) |
|---|---|
| oranges | 1.15 |
| grapes | 0.95 |

F. $2.66          H. $6.57

G. $3.91          I. $10.40

10. **SHORT RESPONSE** On Monday, Jordan biked $2\frac{3}{4}$ miles. On Wednesday, he biked twice as many miles than he did on Monday. On Friday, Jordan biked $8\frac{1}{4}$ miles. How many times more miles did he bike on Friday than he did on Wednesday?

11. Albert used $\frac{3}{8}$ of a half-gallon of paint. What fraction of a gallon of paint did he use?

A. $\frac{3}{4}$     B. $\frac{2}{5}$     C. $\frac{3}{16}$     D. $1\frac{1}{3}$

12. Courtney found the area of her notebook paper shown. What is the width of the piece of notebook paper?

F. 8 inches

G. 8.5 inches

H. 9 inches

I. 9.5 inches

$A = 93.5$ sq. in.

11 in.

13. **SHORT RESPONSE** Lonnie is using the recipe below to make fruit punch. He wants to use $4\frac{1}{2}$ cups of pineapple juice. How many cups of orange juice should he use?

Fruit Punch

$\frac{3}{4}$ cup pineapple juice

$1\frac{1}{2}$ cups orange juice

$\frac{1}{4}$ cup lime juice

14. **EXTENDED RESPONSE** Rose's recipe for snack mix is shown. The recipe serves 12 people but Rose wants to make enough for 30 people.

| Snack Mix Recipe | |
|---|---|
| Ingredients | Amount (cups) |
| Pretzels | $1\frac{1}{2}$ |
| Dry Cereal | $2\frac{3}{4}$ |
| Peanuts | $1\frac{1}{8}$ |
| Margarine | $\frac{1}{3}$ |
| Soy Sauce | $\frac{5}{32}$ |

*Part A* Explain how Rose can calculate the amount of each ingredient she needs to serve 30 people.

*Part B* How much of each ingredient will Rose need?

*Part C* How many cups of dry snack mix will there be in all? Explain your answer.

# Student Recording Sheet

*Use this recording sheet with the Standardized Test Practice pages.*

**Fill in the correct answer. For gridded-response questions, write your answers in the boxes on the answer grid and fill in the bubbles to match your answers.**

1. Ⓐ Ⓑ Ⓒ Ⓓ

2. Ⓕ Ⓖ Ⓗ Ⓘ

3. Ⓐ Ⓑ Ⓒ Ⓓ

4.

5.

6. Ⓕ Ⓖ Ⓗ Ⓘ

7. Ⓐ Ⓑ Ⓒ Ⓓ

8.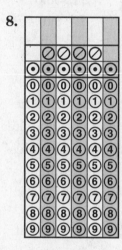

9. Ⓕ Ⓖ Ⓗ Ⓘ

10. _____

    _____

11. Ⓐ Ⓑ Ⓒ Ⓓ

12. Ⓕ Ⓖ Ⓗ Ⓘ

13. _____

## Extended Response

Record your answers for Exercise 14 on the back of this paper.

# Extended-Response Test

**Demonstrate your knowledge by giving a clear, concise solution to each problem. Be sure to include all relevant drawings and justify your answers. You may show your solution in more than one way or investigate beyond the requirements of the problem. If necessary, record your answer on another piece of paper.**

1. You are planning to make a collage and are gathering materials. For the backing you will use a piece of heavy recycled cardboard that is $4\frac{3}{4}$ feet long and $2\frac{2}{5}$ feet wide. One side of the cardboard will be covered with fabric. The other side will be left bare against the wall when you hang the collage.

   a. Estimate how much fabric you will need. Explain how you found your estimate.

   b. Find the actual amount of fabric you will need. Show all your work and explain each step, so someone who doesn't know how to work the problem will understand.

2. In your collage, you want to scatter a few colorful plastic rectangles that are each $\frac{15}{16}$ inch long and $\frac{1}{4}$ inch wide.

   a. Estimate the area of a plastic rectangle. Explain how you found your answer.

   b. Find the actual area. Show your work.

3. You decide to place a design in the middle of the cardboard. This will take up an area of $4\frac{1}{2}$ square feet. Rectangular cut outs that have an area of $\frac{3}{8}$ square feet will cover this area. How many rectangles will fit in the middle area? Explain how you found your answer.

# Extended-Response Rubric

| Score | Description |
|:---:|:---|
| 4 | A score of four is a response in which the student demonstrates a thorough understanding of the mathematics concepts and/or procedures embodied in the task. The student has responded correctly to the task, used mathematically sound procedures, and provided clear and complete explanations and interpretations.<br><br>The response may contain minor flaws that do not detract from the demonstration of a thorough understanding. |
| 3 | A score of three is a response in which the student demonstrates an understanding of the mathematics concepts and/or procedures embodied in the task. The student's response to the task is essentially correct with the mathematical procedures used and the explanations and interpretations provided demonstrating an essential but less than thorough understanding.<br><br>The response may contain minor flaws that reflect inattentive execution of mathematical procedures or indications of some misunderstanding of the underlying mathematics concepts and/or procedures. |
| 2 | A score of two indicates that the student has demonstrated only a partial understanding of the mathematics concepts and/or procedures embodied in the task. Although the student may have used the correct approach to obtaining a solution or may have provided a correct solution, the student's work lacks an essential understanding of the underlying mathematical concepts.<br><br>The response contains errors related to misunderstanding important aspects of the task, misuse of mathematical procedures, or faulty interpretations of results. |
| 1 | A score of one indicates that the student has demonstrated a very limited understanding of the mathematics concepts and/or procedures embodied in the task. The student's response is incomplete and exhibits many flaws. Although the student's response has addressed some of the conditions of the task, the student reached an inadequate conclusion and/or provided reasoning that was faulty or incomplete.<br><br>The response exhibits many flaws or may be incomplete. |
| 0 | A score of zero indicates that the student has provided no response at all, or a completely incorrect or uninterpretable response, or demonstrated insufficient understanding of the mathematics concepts and/or procedures embodied in the task. For example, a student may provide some work that is mathematically correct, but the work does not demonstrate even a rudimentary understanding of the primary focus of the task. |

# Test, Form 1A

Write the letter for the correct answer in the blank at the right of each question.

**Which is the best estimate of each product?**

1. $\frac{1}{5} \times 26$

   **A.** 26     **B.** 21     **C.** 5     **D.** 1

   1. _____

2. $\frac{11}{12} \times \frac{4}{5}$

   **F.** 0     **G.** 1     **H.** 2     **I.** 4

   2. _____

3. $3\frac{1}{3} \times \frac{5}{6}$

   **A.** 4     **B.** 3     **C.** 2     **D.** 1

   3. _____

4. Joelle's necklace is $10\frac{1}{4}$ inches long. Erin's necklace is $2\frac{2}{3}$ times as long. About how long is Erin's necklace?

   **F.** 33 in.     **G.** 30 in.     **H.** 22 in.     **I.** 20 in.

   4. _____

**What is the value of each expression in simplest form?**

5. $4 \times \frac{1}{8}$

   **A.** 4     **B.** 2     **C.** 1     **D.** $\frac{1}{2}$

   5. _____

6. $\frac{1}{3} \times \frac{1}{6}$

   **F.** $\frac{1}{18}$     **G.** $\frac{1}{9}$     **H.** $\frac{1}{2}$     **I.** 2

   6. _____

7. $2\frac{1}{2} \times 1\frac{1}{2}$

   **A.** 4     **B.** $3\frac{3}{4}$     **C.** 3     **D.** $2\frac{1}{4}$

   7. _____

8. What is the area of a room that is $3\frac{3}{4}$ yards long by $3\frac{1}{3}$ yards wide?

   **F.** $12\frac{1}{2}$ yd²     **G.** $10\frac{1}{2}$ yd²     **H.** $9\frac{1}{4}$ yd²     **I.** 5 yd²

   8. _____

9. Olivia ate $\frac{1}{4}$ of a pizza. If there were 12 slices of pizza, how many slices did Olivia eat?

   **A.** 2 slices     **B.** 3 slices     **C.** 4 slices     **D.** 5 slices

   9. _____

10. What is the area of a rectangle with a length of $\frac{1}{3}$ yard and width of $\frac{3}{4}$ yard?

    **F.** $\frac{1}{2}$ yd²     **G.** $\frac{1}{12}$ yd²     **H.** $\frac{1}{4}$ yd²     **I.** $\frac{1}{6}$ yd²

    10. _____

# Test, Form 1A   (continued)

11. Use the *draw a diagram* strategy to solve. Lukas used $\frac{3}{4}$ of the nails in a box. He has 12 nails left. How many did he use?

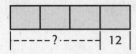

**A.** 21      **B.** 24      **C.** 36      **D.** 48      11. _____

12. Norah has $\frac{2}{3}$ ton of stone to spread equally in 4 square yards. How many tons of stone will be spread in each square yard?

**F.** $\frac{1}{2}$ ton      **G.** 1 ton      **H.** $\frac{1}{6}$ ton      **I.** $2\frac{2}{3}$ tons      12. _____

**What is the value of each expression in simplest form?**

13. $\frac{3}{4} \div \frac{1}{8}$

**A.** $\frac{1}{2}$      **B.** 8      **C.** $\frac{3}{32}$      **D.** 6      13. _____

14. $2 \div \frac{4}{5}$

**F.** $3\frac{1}{2}$      **G.** $2\frac{1}{2}$      **H.** $1\frac{3}{8}$      **I.** $1\frac{1}{5}$      14. _____

15. $3 \div 1\frac{1}{4}$

**A.** $3\frac{3}{4}$      **B.** $3\frac{1}{4}$      **C.** $2\frac{2}{5}$      **D.** $1\frac{2}{5}$      15. _____

16. $1\frac{1}{3} \div \frac{2}{3}$

**F.** $2\frac{2}{3}$      **G.** 2      **H.** $\frac{8}{9}$      **I.** $\frac{1}{2}$      16. _____

17. $4\frac{1}{6} \div 1\frac{2}{3}$

**A.** $\frac{2}{5}$      **B.** $2\frac{1}{2}$      **C.** $4\frac{1}{4}$      **D.** $6\frac{17}{18}$      17. _____

**Complete.**

18. $4\frac{2}{3}$ yd = _____ ft

**F.** 10      **G.** 12      **H.** $12\frac{2}{3}$      **I.** 14      18. _____

19. 68 oz = _____ lb

**A.** $4\frac{1}{3}$      **B.** $4\frac{1}{4}$      **C.** $4\frac{2}{3}$      **D.** $4\frac{3}{4}$      19. _____

20. Addison made 11 quarts of punch for her party. How many gallons of punch did she make?

**F.** 2 gal      **G.** 2.25 gal      **H.** 2.75 gal      **I.** 3 gal      20. _____

# Test, Form 1B

Write the letter for the correct answer in the blank at the right of each question.

**Which is the best estimate for each product?**

1. $\frac{1}{3} \times 20$

   **A.** 20     **B.** 21     **C.** 7     **D.** 3        1. _____

2. $\frac{1}{6} \times \frac{7}{8}$

   **F.** 2     **G.** 1     **H.** $\frac{1}{2}$     **I.** 0        2. _____

3. $2\frac{2}{3} \times 3\frac{1}{4}$

   **A.** 12     **B.** 9     **C.** 6     **D.** 0        3. _____

4. Leyla played sports for $3\frac{3}{4}$ hours. Puno's time playing sports was $\frac{3}{8}$ as long. About how many hours did Puno play sports?

   **F.** 4 h     **G.** 3 h     **H.** 2 h     **I.** 1 h        4. _____

**What is the value of each expression in simplest form?**

5. $5 \times \frac{1}{6}$

   **A.** 30     **B.** 5     **C.** $\frac{6}{5}$     **D.** $\frac{5}{6}$        5. _____

6. $\frac{1}{4} \times \frac{4}{5}$

   **F.** 20     **G.** 5     **H.** $\frac{1}{4}$     **I.** $\frac{1}{5}$        6. _____

7. $3\frac{1}{2} \times 1\frac{1}{2}$

   **A.** $5\frac{1}{4}$     **B.** $4\frac{1}{2}$     **C.** 4     **D.** $3\frac{3}{4}$        7. _____

8. An envelope is $3\frac{1}{3}$ inches long by $3\frac{1}{2}$ inches wide. What is the area of the envelope?

   **F.** 12 in²     **G.** $11\frac{2}{3}$ in²     **H.** $10\frac{1}{2}$ in²     **I.** $9\frac{1}{6}$ in²        8. _____

9. Justin ate $\frac{1}{4}$ of a pie. If there were 8 slices of pie, how many slices did Justin eat?

   **A.** 2 slices     **B.** 3 slices     **C.** 4 slices     **D.** 5 slices        9. _____

10. What is the area of a rectangle with a length of $\frac{3}{4}$ yard and width of $\frac{5}{6}$ yard?

   **F.** $\frac{2}{3}$ yd²     **G.** $\frac{5}{12}$ yd²     **H.** $\frac{1}{2}$ yd²     **I.** $\frac{5}{8}$ yd²        10. _____

# Test, Form 1B  *(continued)*

11. Use the *draw a diagram* strategy to solve. Gino used $\frac{3}{4}$ of the nails in a box. He has 16 nails left. How many did he use?

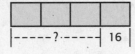

    **A.** 64        **B.** 48        **C.** 32        **D.** 19      11. _____

12. Jason has $\frac{3}{4}$ ton of stone to spread equally in 3 square yards. How many tons of stone will be spread in each square yard?

    **F.** $\frac{1}{4}$ ton      **G.** $\frac{1}{2}$ ton      **H.** $\frac{1}{6}$ ton      **I.** 1 ton      12. _____

**What is the value of each expression in simplest form?**

13. $\frac{1}{2} \div \frac{1}{3}$

    **A.** $\frac{1}{6}$        **B.** $\frac{1}{2}$        **C.** 3        **D.** $1\frac{1}{2}$      13. _____

14. $3 \div \frac{5}{6}$

    **F.** $\frac{5}{18}$        **G.** $2\frac{1}{2}$        **H.** 3        **I.** $3\frac{3}{5}$      14. _____

15. $2 \div 1\frac{2}{3}$

    **A.** $\frac{5}{6}$        **B.** $1\frac{1}{5}$        **C.** $1\frac{1}{3}$        **D.** $3\frac{1}{3}$      15. _____

16. $2\frac{1}{2} \div \frac{1}{2}$

    **F.** 5        **G.** 2        **H.** $1\frac{1}{4}$        **I.** $\frac{4}{5}$      16. _____

17. $3\frac{1}{4} \div 4\frac{1}{3}$

    **A.** $14\frac{1}{2}$        **B.** $1\frac{5}{8}$        **C.** $1\frac{1}{3}$        **D.** $\frac{3}{4}$      17. _____

**Complete.**

18. $6\frac{1}{2}$ qt = _____ pt

    **F.** 13        **G.** 26        **H.** $3\frac{1}{4}$        **I.** $1\frac{5}{8}$      18. _____

19. $2\frac{1}{4}$ T = _____ lb

    **A.** 3,500      **B.** 4,500      **C.** 4,750      **D.** 4,800      19. _____

20. Rylie made 18 cups of punch for her party. How many fluid ounces of punch did she make?

    **F.** 144 fl oz      **G.** 145 fl oz      **H.** 4.5 fl oz      **I.** 2.25 fl oz      20. _____

# Test, Form 2A

**Write the letter for the correct answer in the blank at the right of each question.**

**Which is the best estimate for each product?**

1. $27 \times \frac{1}{7}$

   **A.** 2          **B.** 4          **C.** 7          **D.** 10                    1. _____

2. $\frac{9}{10} \times \frac{5}{11}$

   **F.** 2          **G.** 1          **H.** $\frac{1}{2}$          **I.** 0                    2. _____

3. It took Morry $2\frac{5}{6}$ hours to drive to the beach. It took Jason twice as long. About how long did it take Jason to drive to the beach?

   **A.** 1 hour          **B.** 3 hours          **C.** 4 hours          **D.** 6 hours          3. _____

**What is the value of each expression in simplest form?**

4. $12 \times \frac{1}{8}$

   **F.** 12          **G.** 8          **H.** $1\frac{1}{2}$          **I.** $\frac{3}{4}$                    4. _____

5. $\frac{5}{6} \times \frac{7}{10}$

   **A.** $\frac{7}{12}$          **B.** $\frac{35}{60}$          **C.** $1\frac{4}{21}$          **D.** $\frac{50}{42}$          5. _____

6. $1\frac{2}{3} \times 2\frac{3}{5}$

   **F.** $4\frac{1}{3}$          **G.** $4\frac{2}{3}$          **H.** $4\frac{2}{5}$          **I.** $2\frac{2}{5}$                    6. _____

7. Ms. Liang is building a deck that is $2\frac{2}{9}$ yards long and $3\frac{2}{5}$ yards wide. What is the area of her deck?

   **A.** $6\frac{4}{5}$ yd²          **B.** $6\frac{3}{5}$ yd²          **C.** $7\frac{5}{9}$ yd²          **D.** $7\frac{2}{15}$ yd²          7. _____

8. Use the *draw a diagram* strategy to solve. Leon read $\frac{5}{7}$ of the pages in his book. He has 28 pages left to read. How many pages did he read already?

   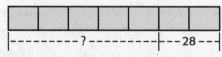

   **F.** 70          **G.** 98          **H.** 140          **I.** 196          8. _____

9. John's dog slept for 3 hours. If the dog snored every $\frac{1}{3}$ of an hour, how many times did he snore?

   **A.** 1 time          **B.** 3 times          **C.** 6 times          **D.** 9 times          9. _____

# Test, Form 2A   *(continued)*

**What is the value of each expression in simplest form?**

**10.** $8 \div \frac{1}{3}$

    **F.** $\frac{8}{3}$         **G.** $2\frac{1}{3}$         **H.** 12         **I.** 24         **10.** _____

**11.** $9 \div \frac{3}{5}$

    **A.** 15         **B.** $5\frac{2}{5}$         **C.** $4\frac{2}{3}$         **D.** $2\frac{2}{5}$         **11.** _____

**12.** $\frac{1}{2} \div \frac{2}{3}$

    **F.** $\frac{5}{6}$         **G.** $\frac{1}{3}$         **H.** $\frac{3}{4}$        **I.** $1\frac{1}{2}$         **12.** _____

**13.** $\frac{7}{8} \div \frac{3}{4}$

    **A.** $1\frac{1}{6}$         **B.** $1\frac{2}{7}$         **C.** $1\frac{1}{8}$         **D.** $\frac{21}{32}$         **13.** _____

**14.** $3\frac{1}{3} \div 1\frac{7}{8}$

    **F.** $6\frac{1}{4}$         **G.** $\frac{29}{8}$         **H.** $1\frac{7}{9}$         **I.** $\frac{9}{16}$         **14.** _____

**15.** Brandon has $7\frac{1}{2}$ gallons of paint. He plans on using $1\frac{1}{2}$ gallons on each room. How many rooms will he be able to paint?     **15.** _____

**16.** Find $\frac{4}{5} \times 1\frac{2}{3}$.     **16.** _____

**17.** Find $3\frac{2}{5} \div 1\frac{1}{10}$.     **17.** _____

**18.** Eight bricks are laid end to end along the edge of a flower bed. Each brick is $8\frac{1}{2}$ inches long. How long is the row of bricks, in feet?     **18.** _____

**19.** A box contains $1\frac{4}{5}$ pounds of pasta. How many ounces of pasta are in the box?     **19.** _____

**20.** The area of a bathroom floor measures 2 yards by 3 yards and each custom tile that makes up the flooring is $1\frac{1}{2}$ square feet. How many tiles are needed to cover the floor?     **20.** _____

# Test, Form 2B

**Write the letter for the correct answer in the blank at the right of each question.**

**Which is the best estimate for each product?**

**1.** $31 \times \frac{1}{8}$

   **A.** 10         **B.** 7         **C.** 4         **D.** 3         **1.** _____

**2.** $\frac{11}{12} \times \frac{7}{8}$

   **F.** 2         **G.** 1         **H.** $\frac{1}{2}$         **I.** 0         **2.** _____

**3.** Mia is making costumes for a play. Each costume needs $3\frac{7}{9}$ yards of velvet. She is making 6 costumes. About how much velvet does she need?

   **3.** _____

   **A.** 12 yards    **B.** 18 yards    **C.** 24 yards    **D.** 30 yards

**What is the value of each expression in simplest form?**

**4.** $15 \times \frac{1}{6}$

   **F.** 15         **G.** 12         **H.** $3\frac{1}{2}$         **I.** $2\frac{1}{2}$         **4.** _____

**5.** $\frac{5}{8} \times \frac{8}{9}$

   **A.** $1\frac{4}{5}$         **B.** $\frac{45}{64}$         **C.** $\frac{40}{72}$         **D.** $\frac{5}{9}$         **5.** _____

**6.** $1\frac{3}{5} \times 3\frac{1}{2}$

   **F.** $3\frac{3}{10}$         **G.** $5\frac{3}{5}$         **H.** $5\frac{1}{2}$         **I.** $3\frac{3}{5}$         **6.** _____

**7.** Ms. Oliver sewed together a quilt that is $3\frac{2}{3}$ feet long and $2\frac{1}{4}$ feet wide. What is the area of her quilt?

   **A.** $5\frac{1}{2}$ ft²    **B.** $6\frac{1}{6}$ ft²    **C.** $8\frac{1}{4}$ ft²    **D.** 9 ft²    **7.** _____

**8.** Use the *draw a diagram* strategy to solve. Corbin read $\frac{4}{7}$ of the pages in his book. He has 42 pages left to read. How many pages did he read already?

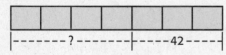

   **F.** 56         **G.** 84         **H.** 98         **I.** 147         **8.** _____

**9.** Joe has 12 cups of soup in a pot. If he pours $\frac{1}{4}$ cup of soup into each bowl, how many bowls will be used?

   **A.** 48 bowls    **B.** 36 bowls    **C.** 16 bowls    **D.** 3 bowls    **9.** _____

# Test, Form 2B   (continued)

**What is the value of each expression in simplest form?**

**10.** $10 \div \frac{1}{6}$

     **F.** 60      **G.** 16      **H.** $1\frac{2}{3}$      **I.** $\frac{3}{5}$      **10.** _____

**11.** $7 \div \frac{3}{4}$

     **A.** $\frac{4}{21}$      **B.** $5\frac{1}{4}$      **C.** $9\frac{1}{3}$      **D.** 28      **11.** _____

**12.** $7 \div 1\frac{1}{2}$

     **F.** 14      **G.** $10\frac{1}{2}$      **H.** $4\frac{2}{3}$      **I.** $\frac{3}{14}$      **12.** _____

**13.** $2\frac{5}{8} \div \frac{3}{4}$

     **A.** $1\frac{31}{32}$      **B.** $2\frac{5}{6}$      **C.** $3\frac{1}{3}$      **D.** $3\frac{1}{2}$      **13.** _____

**14.** $6\frac{3}{4} \div 2\frac{1}{4}$

     **F.** $3\frac{1}{4}$      **G.** 3      **H.** $\frac{9}{4}$      **I.** $\frac{4}{9}$      **14.** _____

**15.** Lyndsay has $3\frac{3}{4}$ gallons of paint. She plans on using $1\frac{1}{4}$ gallons on each room. How many rooms will she be able to paint?      **15.** _____

**16.** Find $\frac{2}{7} \times 3\frac{1}{2}$.      **16.** _____

**17.** Find $4\frac{1}{2} \div 1\frac{5}{8}$.      **17.** _____

**18.** Six pipes are laid end to end to form a row. Each pipe is $9\frac{1}{3}$ inches long. How long is the row of pipes, in feet?      **18.** _____

**19.** A box contains $1\frac{3}{4}$ pounds of beans. How many ounces of beans are in the box?      **19.** _____

**20.** The area of a closet floor measures 3 yards by 3 yards and each custom tile that makes up the flooring is $1\frac{1}{2}$ square feet. How many tiles are needed to cover the floor?      **20.** _____

# Test, Form 3A

SCORE _____

**Estimate each product.**

1. $\frac{1}{9} \times 39$

1. _____

2. $\frac{10}{11} \times \frac{3}{5}$

2. _____

3. The average population of a certain city is $4\frac{7}{8}$ million people. Suppose each person produces an average of $2\frac{2}{9}$ pounds of garbage each day. About how much garbage would be produced each day?

3. _____

**Multiply. Write in simplest form.**

4. $8 \times \frac{4}{9}$

4. _____

5. $\frac{3}{16} \times \frac{5}{12}$

5. _____

6. $\frac{3}{7} \times 2\frac{5}{8}$

6. _____

7. A container holds $\frac{1}{4}$ gallon of oil. Jonathan is changing the oil on his car and needs 5 containers. How many gallons of oil does Jonathan need?

7. _____

8. Renee spent $\frac{1}{8}$ of the day landscaping around the house. She spent $\frac{1}{2}$ of that time pulling weeds. What fraction of the day did Renee spend pulling weeds?

8. _____

9. A quilt measures $4\frac{2}{3}$ feet by 6 feet. What is the area of the quilt?

9. _____

10. A recipe make $5\frac{1}{2}$ dozen cookies. Marquis needs to make $3\frac{3}{4}$ times this amount. How many dozens of cookies will he make?

10. _____

11. Use the *draw a diagram* strategy to solve. Elvin read $\frac{3}{7}$ of the pages in his book. He has 56 pages left to read. How many pages has he read already?

11. _____

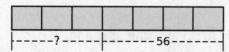

# Test, Form 3A  (continued)

**Divide. Write in simplest form.**

12. $5 \div \frac{1}{7}$

12. _____

13. $\frac{1}{2} \div \frac{5}{9}$

13. _____

14. $\frac{2}{3} \div \frac{5}{7}$

14. _____

15. $\frac{7}{9} \div 5\frac{4}{9}$

15. _____

**Solve each problem. Write in simplest form.**

16. The pilot on the flight that Katelyn was on announced that they were flying at 29,040 feet. Was Katelyn over 6 miles above the ground? Explain.

16. _____

17. Spencer spent 4 hours doing chores over the weekend. If he spent $\frac{2}{3}$ of an hour for each chore, how many chores did he do?

17. _____

18. Duane bought $68\frac{3}{4}$ inches of chain for an art project. How many 15-inch chains can he make from it?

18. _____

19. A cooler contains $13\frac{1}{2}$ cups of fruit juice. How many pints of fruit juice does the cooler contain?

19. _____

20. Mrs. Franks has 45 pounds of grain to divide among the farm animals. If each animal receives $\frac{3}{4}$ pound of grain for each feeding, how many times can Mrs. Franks feed her animals?

20. _____

# Test, Form 3B

**Estimate each product.**

1. $\frac{1}{11} \times 52$

         1. _____

2. $\frac{7}{15} \times 1\frac{7}{9}$

         2. _____

3. A van is traveling at $64\frac{7}{8}$ miles per hour. An airplane is traveling $5\frac{6}{7}$ times faster. About how fast is the airplane traveling?

         3. _____

**Multiply. Write in simplest form.**

4. $11 \times \frac{9}{10}$

         4. _____

5. $\frac{2}{5} \times \frac{3}{4}$

         5. _____

6. $\frac{3}{8} \times 3\frac{1}{6}$

         6. _____

7. A container holds $\frac{3}{4}$ gallon of oil. Joan is changing the oil on his vehicles and needs 6 containers. How many gallons of oil does Joan need?

         7. _____

8. Natasha spent $\frac{1}{4}$ of the day cleaning the kitchen. She spent $\frac{1}{3}$ of that time washing dishes. What fraction of the day did Natasha spend washing dishes?

         8. _____

9. A certain type of fencing sells for $9 a yard. How much will $10\frac{2}{3}$ yards of fencing cost?

         9. _____

10. Phillip rides his scooter to school in $6\frac{2}{5}$ minutes. If it takes Josef $4\frac{1}{2}$ times as long to get to school, how long does it take Josef to get to school?

         10. _____

11. Use the *draw a diagram* strategy to solve. Elvin read $\frac{4}{7}$ of the pages in his book. He has 126 pages left to read. How many pages has he read already?

         11. _____

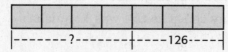

**Course 1 • Chapter 4** Multiply and Divide Fractions

# Test, Form 3B  (continued)

**Divide. Write in simplest form.**

12. $9 \div \frac{1}{7}$

12. _____

13. $\frac{1}{4} \div \frac{3}{10}$

13. _____

14. $\frac{5}{6} \div \frac{1}{8}$

14. _____

15. $\frac{3}{4} \div 5\frac{1}{3}$

15. _____

**Solve each problem. Write in simplest form.**

16. Mr. Lash has 42 pounds of dog food to divide among his dogs. If each dog receives $\frac{7}{8}$ pound of food for each feeding, how many times can Mr. Lash feed his animals?

16. _____

17. The pilot on the flight that Morgan was on announced that they were flying at 30,360 feet. Was Morgan over 6 miles above the ground? Explain.

17. _____

18. Jose has 21 pounds of clay. He needs $5\frac{1}{4}$ pounds of clay to make one piece of pottery. How many pieces of pottery can he make?

18. _____

19. A cooler contains $14\frac{1}{2}$ cups of juice. How many pints of juice does the cooler contain?

19. _____

20. Ben made a model that was $5\frac{3}{4}$ inches tall. Laquisha made a model that was $70\frac{11}{12}$ inches tall. How many times as tall was Laquisha's model?

20. _____

**Course 1 • Chapter 4** Multiply and Divide Fractions

# Are You Ready?

## Review

### Example 1

Find $\frac{5}{7} + \frac{2}{3}$. Write in simplest form.

$\frac{5}{7} + \frac{2}{3} = \frac{5 \times 3}{7 \times 3} + \frac{2 \times 7}{3 \times 7}$    Rename using the LCD, 21.

$\qquad = \frac{15}{21} + \frac{14}{21}$    Add the fractions.

$\qquad = \frac{29}{21}$    Simplify.

$\qquad = 1\frac{8}{21}$    Rewrite as a mixed number.

### Example 2

Find $\frac{4}{9} - \frac{1}{3}$. Write in simplest form.

$\frac{4}{9} - \frac{1}{3} = \frac{4}{9} - \frac{1 \times 3}{3 \times 3}$    Rename using the LCD, 9.

$\qquad = \frac{4}{9} - \frac{3}{9}$    Subtract the fractions.

$\qquad = \frac{1}{9}$    Simplify.

**Add or subtract. Write in simplest form.**

1. $\frac{3}{4} + \frac{1}{5}$    2. $\frac{2}{5} - \frac{1}{4}$

3. $\frac{8}{9} - \frac{1}{3}$    4. $\frac{2}{7} + \frac{4}{5}$

5. $\frac{7}{8} + \frac{1}{2}$    6. $\frac{5}{6} - \frac{1}{4}$

7. $\frac{7}{9} - \frac{2}{5}$    8. $\frac{6}{7} + \frac{7}{8}$

9. $\frac{4}{5} + \frac{1}{9}$    10. $\frac{3}{8} - \frac{1}{7}$

1. _____

2. _____

3. _____

4. _____

5. _____

6. _____

7. _____

8. _____

9. _____

10. _____

# Are You Ready?

## Practice

**Add or subtract. Write in simplest form.**

1. $\dfrac{2}{3} + \dfrac{1}{8}$

1. _____

2. $\dfrac{3}{5} + \dfrac{3}{4}$

2. _____

3. $\dfrac{7}{9} - \dfrac{2}{3}$

3. _____

4. $\dfrac{6}{7} - \dfrac{2}{5}$

4. _____

5. $\dfrac{8}{9} + \dfrac{1}{5}$

5. _____

6. $\dfrac{4}{5} - \dfrac{1}{7}$

6. _____

7. **SCHOOL** One eighth of the sixth-grade class chose science as their favorite subject. Two fifths chose social studies. How much more of the class chose social studies than science?

7. _____

**Multiply or divide. Write in simplest form.**

8. $\dfrac{4}{5} \times \dfrac{2}{7}$

8. _____

9. $\dfrac{3}{4} \times \dfrac{7}{9}$

9. _____

10. $\dfrac{1}{7} \div \dfrac{5}{6}$

10. _____

11. $\dfrac{7}{8} \div \dfrac{1}{2}$

11. _____

12. $\dfrac{2}{3} \times \dfrac{4}{9}$

12. _____

13. $\dfrac{5}{8} \div \dfrac{3}{5}$

13. _____

14. **TREATS** A teacher had $\dfrac{3}{4}$ of a bag of treats for her students. If she gave them $\dfrac{1}{2}$ of the treats she had, what fraction of the entire bag of treats did she give the students?

14. _____

# Are You Ready?

## Apply

1. **ATHLETICS** Two thirds of the track team ran laps after practice. One fourth of the team worked out in the weight room. What fraction more of the team ran laps than worked out in the weight room?

2. **ALLOWANCE** Eustace used $\frac{1}{2}$ of his allowance to buy CDs. He put $\frac{1}{3}$ of his allowance in savings. What fraction of his allowance did he use for CDs and savings?

3. **PAPER** Arcus had $\frac{8}{9}$ of a pack of construction paper. He used $\frac{1}{5}$ of the paper to make paper airplanes. What fraction of the original pack did Arcus use to make paper airplanes?

4. **MEASUREMENT** Lagan sold $\frac{7}{8}$ pound of chocolate for the school fundraiser. Nathan sold $\frac{6}{7}$ pound. How much more did Lagan sell than Nathan?

5. **CONSTRUCTION** A carpenter bought $\frac{1}{2}$ pound of size 8 nails and $\frac{3}{8}$ pound of size 16 nails. How many pounds of nails did he buy in all?

6. **FARMING** A farmer planted $\frac{4}{5}$ of a field in soybeans. One day, $\frac{5}{6}$ of the soybeans were picked. What fraction of the entire field was picked that day?

# Diagnostic Test

**Replace ● with <, >, or = to make a true statement.**

1. 468,523 ● 648,523

2. 895,000 ● 89,500

3. 36,542 ● 38,542

4. 972,314 ● 972,413

5. **TRUCKING** A trucker drove 214 miles one day and 241 miles the next day. Compare 214 and 241 using <, >, or =.

6. **COMPANY** A computer company sold 2,380 new computers last year. This year, they sold 2,208 new computers. Compare 2,380 and 2,208 using <, >, or =.

**Write each fraction as a decimal.**

7. $\frac{7}{8}$

8. $\frac{13}{20}$

9. $\frac{4}{25}$

10. $\frac{37}{50}$

1. _____

2. _____

3. _____

4. _____

5. _____

6. _____

7. _____

8. _____

9. _____

10. _____

# Pretest

**Write an integer for each situation.**

1. $50 withdrawal

2. elevation of 145 feet

**Evaluate each expression.**

3. $|30 - 5| + |-7|$

4. $|-2| + |10 + 2|$

5. **TEMPERATURE** The low temperature in one city was −4°F. The low temperature in another city was 8°F. Write an inequality to compare the temperatures.

**Write each fraction as a decimal. Use bar notation if necessary.**

6. $\frac{4}{9}$

7. $\frac{8}{15}$

8. $-\frac{11}{12}$

**Graph and label each point on the coordinate plane to the right.**

9. $(-4, -2)$

10. $(4, -3)$

11. $(-1, 1)$

12. $(2, 4)$

1. _____

2. _____

3. _____

4. _____

5. _____

6. _____

7. _____

8. _____

9–12.

# Chapter Quiz

**Write an integer for each situation.**

1. a loss of 14 pounds

2. rising 7 degrees

3. 35 feet below sea level

**Graph each set of integers on the number line.**

4. {−2, 3, 0, −1}

5. {2, −4, 5, −3}

**Evaluate each expression.**

6. $|-46|$

7. $|6| - |-2|$

8. $|-5| + |9 + 3|$

9. Manuel is hiking up to a mountain top that is 1,207 feet above sea level. Stephanie scuba dived to 23 feet below sea level. What is the difference between these two measurements?

**Replace each ● with <, >, or = to make a true statement.**

10. −7 ● 3

11. 10 ● −1

12. 0 ● −2

13. 18 ● −19

**Order each set of integers from least to greatest.**

14. 7, −6, 4, −10, 3

15. 16, −24, 9, −30, 25

16. 93, −76, 102, −101, 83, −48

1. _____

2. _____

3. _____

4.
−5−4−3−2−1 0 1 2 3 4 5

5.
−5−4−3−2−1 0 1 2 3 4 5

6. _____

7. _____

8. _____

9. _____

10. _____

11. _____

12. _____

13. _____

14. _____

15. _____

16. _____

# Vocabulary Test

| | | |
|---|---|---|
| absolute value | opposites | repeating decimal |
| bar notation | positive integer | terminating decimal |
| integer | quadrants | |
| negative integer | rational number | |

**Choose from the terms above to complete each sentence.**

1. The absolute value of a negative integer is a _____.

   1. _____

2. The integers −10 and 10 are _____.

   2. _____

3. The _____ of −5 is 5.

   3. _____

4. Any number that can be written as a fraction is a(n) _____.

   4. _____

5. Repeating decimals can be expressed exactly using _____.

   5. _____

6. A(n) _____ is a decimal whose division ends.

   6. _____

7. 0.8585858585... is an example of a(n) _____.

   7. _____

**Define each term in your own words.**

8. quadrants

   8. _____

9. integer

   9. _____

# Standardized Test Practice

**Read each question. Then fill in the correct answer on the answer sheet provided by your teacher or on a sheet of paper.**

1. ≡≡✍ **GRIDDED RESPONSE** The Music Shop records the number of CDs sold for 5 months. What fraction of CDs were sold in April? Write the fraction in simplest form.

| Month | Number of CDs Sold |
|---|---|
| January | 50 |
| February | 35 |
| March | 42 |
| April | 110 |
| May | 98 |

2. Which point on the grid below corresponds to the ordered pair (5,2)?

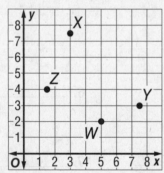

   **A.** Point $W$     **C.** Point $Y$
   **B.** Point $X$     **D.** Point $Z$

3. ≡≡✍ **GRIDDED RESPONSE** The Geography Club is selling wrapping paper for a fundraiser. They ordered 16 cases. If each student in the club takes $\frac{2}{3}$ of a case, how many students are in the club?

4. **SHORT RESPONSE** There are 16 cars and 64 passengers scheduled to go to a concert. Write a ratio in simplest form that compares the number of passengers to the number of cars.

5. Which of the following ordered pairs is located inside the graph of the square?

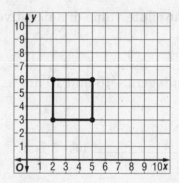

   **F.** (4, 0)     **H.** (2, 2)
   **G.** (1, 6)     **I.** (4, 5)

6. **SHORT RESPONSE** Jacob multiplied 30 by 0.25 and got 7.5. Zane said that it was wrong because when you multiply two numbers, the product is always greater than both of the numbers. Explain who is correct.

7. A container of fruit contains $12\frac{3}{4}$ pounds of oranges, $8\frac{1}{4}$ pounds of bananas, and 7 pounds of apples. What is the total weight, in pounds, of the contents of the container?

   **A.** 27 pounds

   **B.** 28 pounds

   **C.** $27\frac{1}{2}$ pounds

   **D.** $28\frac{1}{2}$ pounds

8. **SHORT RESPONSE** The table shows J.T.'s training schedule for a marathon. If the pattern continues, how many minutes will he run on Day 8?

| Day | Running Time (min) |
|-----|-------------------|
| 1 | 20 |
| 2 | 22 |
| 3 | 24 |
| 4 | 26 |

9. Which point **best** represents the location of the ordered pair $\left(1\frac{1}{4}, 2\right)$?

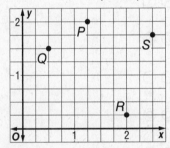

   **F.** Point $P$            **H.** Point $R$
   **G.** Point $Q$            **I.** Point $S$

10. Wilson made punch from a recipe that called for $1\frac{1}{2}$ cups of orange juice and $\frac{3}{4}$ cup lemonade. Wilson used $2\frac{1}{4}$ cups of orange juice and $1\frac{1}{8}$ cups of lemonade. How many batches of punch did he make?

   **A.** 3            **C.** 2
   **B.** $2\frac{1}{2}$            **D.** $1\frac{1}{2}$

11. Dante is training for a cross country meet. He ran 35 miles in 10 days. At this rate, how many miles does Dante run each day?

   **F.** 0.3 mile
   **G.** 3 miles
   **H.** 3.5 miles
   **I.** 4.5 miles

12. The temperature at noon was 8 degrees below zero Celsius. At five o'clock it was 12 degrees below zero Celsius. Which integer represents the temperature at noon in degrees Celsius?

   **A.** −12            **C.** 8
   **B.** −8            **D.** 12

13. Mary's temperature on Sunday night was 101°F. On Monday morning, her temperature was 103°F. Which integer represents Mary's temperature on Sunday night in degrees Fahrenheit?

   **F.** −103            **H.** 101
   **G.** −101            **I.** 103

14. **EXTENDED RESPONSE** The students in Mrs. Steven's homeroom are collecting recycled materials. They collected 3.65 pounds of newspaper, $3\frac{7}{9}$ pounds of glass, and 3.6 pounds of aluminum.

   **Part A** Write the number of pounds of glass as a decimal. Use bar notation if necessary.

   **Part B** Order the three measurements from least to greatest.

# Student Recording Sheet

*Use this recording sheet with the Standardized Test Practice pages.*

**Fill in the correct answer. For gridded-response questions, write your answers in the boxes on the answer grid and fill in the bubbles to match your answers.**

1.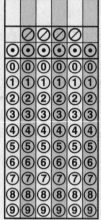

2. Ⓐ Ⓑ Ⓒ Ⓓ

3.

4. _____

5. Ⓕ Ⓖ Ⓗ Ⓘ

6. _____

_____

7. Ⓐ Ⓑ Ⓒ Ⓓ

8. _____

9. Ⓕ Ⓖ Ⓗ Ⓘ

10. Ⓐ Ⓑ Ⓒ Ⓓ

11. Ⓕ Ⓖ Ⓗ Ⓘ

12. Ⓐ Ⓑ Ⓒ Ⓓ

13. Ⓕ Ⓖ Ⓗ Ⓘ

## Extended Response

Record your answers for Exercise 14 on the back of this paper.

# Extended-Response Test

SCORE _____

Demonstrate your knowledge by giving a clear, concise solution to each problem. Be sure to include all relevant drawings and justify your answers. You may show your solution in more than one way or investigate beyond the requirements of the problem. If necessary, record your answer on another piece of paper.

1. A pipe extends 0.875 kilometer below the ground.

   a. In relation to the ground, what is the elevation of the pipe? Write the elevation as a fraction. Explain how you found your answer.

   b. The elevation of an underground river is $2\frac{5}{12}$ kilometers below the ground. What is the elevation of the underground river as a decimal? Explain how you found your answer.

   c. The elevation of the lowest part of a cave is $-3\frac{1}{3}$ kilometers. How does this elevation compare to that of the underground river?

2. The locations of some buildings at the county fair are shown on the coordinate plane.

   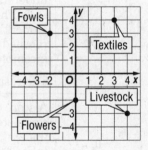

   a. Which ordered pair names the location of the textiles building?

   b. Which ordered pair names the location of the flowers building?

   c. An information booth will be added at $(-3, 2)$. To which building will the information booth be closest?

   d. Three food booths will be added. Describe three possible locations given by an ordered pair in which $y = -1$.

# Extended-Response Rubric

| Score | Description |
|-------|-------------|
| 4 | A score of four is a response in which the student demonstrates a thorough understanding of the mathematics concepts and/or procedures embodied in the task. The student has responded correctly to the task, used mathematically sound procedures, and provided clear and complete explanations and interpretations. |
|  | The response may contain minor flaws that do not detract from the demonstration of a thorough understanding. |
| 3 | A score of three is a response in which the student demonstrates an understanding of the mathematics concepts and/or procedures embodied in the task. The student's response to the task is essentially correct with the mathematical procedures used and the explanations and interpretations provided demonstrating an essential but less than thorough understanding. |
|  | The response may contain minor flaws that reflect inattentive execution of mathematical procedures or indications of some misunderstanding of the underlying mathematics concepts and/or procedures. |
| 2 | A score of two indicates that the student has demonstrated only a partial understanding of the mathematics concepts and/or procedures embodied in the task. Although the student may have used the correct approach to obtaining a solution or may have provided a correct solution, the student's work lacks an essential understanding of the underlying mathematical concepts. |
|  | The response contains errors related to misunderstanding important aspects of the task, misuse of mathematical procedures, or faulty interpretations of results. |
| 1 | A score of one indicates that the student has demonstrated a very limited understanding of the mathematics concepts and/or procedures embodied in the task. The student's response is incomplete and exhibits many flaws. Although the student's response has addressed some of the conditions of the task, the student reached an inadequate conclusion and/or provided reasoning that was faulty or incomplete. |
|  | The response exhibits many flaws or may be incomplete. |
| 0 | A score of zero indicates that the student has provided no response at all, or a completely incorrect or uninterpretable response, or demonstrated insufficient understanding of the mathematics concepts and/or procedures embodied in the task. For example, a student may provide some work that is mathematically correct, but the work does not demonstrate even a rudimentary understanding of the primary focus of the task. |

# Test, Form 1A

**Write the letter for the correct answer in the blank at the right of each question.**

1. Write $-\frac{12}{25}$ as a decimal.

   **A.** $-0.52$  **B.** $-0.48$  **C.** $2.0\overline{8}$  **D.** $12.04$

   1. _____

2. Which of the following is a true statement?

   **F.** $-\frac{5}{6} > -\frac{4}{9}$  **G.** $4.3 > 4\frac{3}{4}$  **H.** $13\frac{5}{8} = 13.625$  **I.** $\frac{5}{9} > 0.\overline{57}$

   2. _____

3. Trent made 11 free throws out of 15 attempts during the basketball game. What was his free-throw average expressed as a decimal?

   **A.** $0.7$  **B.** $0.7\overline{3}$  **C.** $0.3\overline{7}$  **D.** $0.\overline{8}$

   3. _____

4. Which set of rational numbers is ordered from least to greatest?

   **F.** $-4.\overline{3}, 4\frac{1}{4}, 4\frac{1}{5}, 4.06$  **H.** $-6\frac{5}{8}, -6.34, -6.\overline{3}, -6\frac{1}{4}$

   **G.** $-0.27, -\frac{2}{9}, -\frac{2}{3}, -0.\overline{1}$  **I.** $-7\frac{12}{13}, -7\frac{13}{15}, -7.8\overline{6}, -7.\overline{86}$

   4. _____

**Evaluate each expression.**

5. $|-8|$

   **A.** $8$  **B.** $-8$  **C.** $0$  **D.** $4$

   5. _____

6. $|10 - 3|$

   **F.** $-7$  **G.** $10$  **H.** $3$  **I.** $7$

   6. _____

7. $|-14| + |5|$

   **A.** $19$  **B.** $11$  **C.** $-11$  **D.** $19$

   7. _____

8. What is the opposite of $-17$?

   **F.** $-17$  **G.** $17$  **H.** $0$  **I.** $-|17|$

   8. _____

9. Which set of integers is graphed on the number line?

   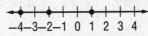

   **A.** $\{-4, -2, -1\}$  **C.** $\{-4, -1, 2\}$

   **B.** $\{-4, -2, 1\}$  **D.** $\{-4, 1, 2\}$

   9. _____

# Test, Form 1A  *(continued)*

**For Exercises 10 and 11, use the coordinate plane below.**

10. Which ordered pair names point *E*?

    **F.** (−1, 5)          **H.** (−5, 1)

    **G.** (1, −5)          **I.** (5, −1)

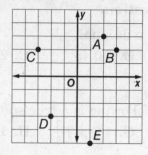

10. _____

11. Which of the following names the point for the ordered pair (2, 3)?

    **A.** point *A*          **C.** point *C*

    **B.** point *B*          **D.** point *D*

11. _____

12. Which situation does the integer −6 best represent?

    **F.** 6 yards behind the winner    **H.** finding $6 in a pocket

    **G.** 6 feet above the ground      **I.** earning $6

12. _____

**Replace each ● with <, >, or = to make a true statement.**

13. 9 ● −8

    **A.** <        **B.** >        **C.** =        **D.** +

13. _____

14. 6 ● −8

    **F.** <        **G.** >        **H.** =        **I.** +

14. _____

15. Which number is less than −3?

    **A.** 0        **B.** 2        **C.** −2        **D.** −4

15. _____

16. Order −5, −7, 0, and 4 from least to greatest.

    **F.** 0, 4, −5, −7          **H.** −7, −5, 0, 4

    **G.** −5, −7, 0, 4          **I.** −7, −5, 4, 0

16. _____

17. Which quadrant contains the point named by (2, 5)?

    **A.** Quadrant I          **C.** Quadrant III

    **B.** Quadrant II         **D.** Quadrant IV

17. _____

18. Kira has $40 to spend and used $20 to buy a new pair of jeans. Which integer best represents the situation of spending $20?

    **F.** 20          **H.** 40

    **G.** −20        **I.** −40

18. _____

19. Geoffrey used 7 out of his 20 tokens on one game at the arcade. What is this fraction written as a decimal?

    **A.** $0.3\overline{5}$        **B.** 0.35        **C.** $0.\overline{4}$        **D.** 0.45

19. _____

# Test, Form 1B

**Write the letter for the correct answer in the blank at the right of each question.**

1. Write $-\frac{18}{25}$ as a decimal.
   **A.** $-0.72$      **B.** $-0.28$      **C.** $1.07$      **D.** $1.3\overline{8}$

   1. _____

2. Which of the following is a true statement?
   **F.** $\frac{5}{11} > \frac{5}{9}$    **G.** $0.\overline{15} < \frac{3}{20}$    **H.** $4\frac{2}{3} > 4\frac{3}{5}$    **I.** $11.\overline{39} < 11\frac{1}{3}$

   2. _____

3. Trent made 7 free throws out of 11 attempts during the basketball game. What was his free-throw average expressed as a decimal?
   **A.** $0.6\overline{3}$      **B.** $0.\overline{63}$      **C.** $0.6$      **D.** $0.7$

   3. _____

4. Which set of rational numbers is ordered from least to greatest?
   **F.** $-4.08, -4\frac{1}{5}, -4\frac{1}{4}, -4.\overline{3}$    **H.** $-0.18, -\frac{3}{10}, -\frac{1}{3}, -0.\overline{2}$

   **G.** $-2\frac{5}{8}, -2.34, -2.\overline{3}, -2\frac{1}{4}$    **I.** $-6\frac{11}{15}, -6\frac{1}{2}, -6.\overline{35}, -6.3\overline{5}$

   4. _____

**Evaluate each expression.**

5. $|-9|$
   **A.** $9$      **B.** $-9$      **C.** $0$      **D.** $5$

   5. _____

6. $|11 + 8|$
   **F.** $-19$      **G.** $19$      **H.** $11$      **I.** $-11$

   6. _____

7. $|10| + |-15|$
   **A.** $10$      **B.** $5$      **C.** $-25$      **D.** $25$

   7. _____

8. What is the opposite of 23?
   **F.** $-23$      **G.** $23$      **H.** $0$      **I.** $|23|$

   8. _____

9. Which set of integers is graphed on the number line?

   ```
   ←+--+--+--+--+--+--●--●--+--+--●--+--+--+--+--●--+→
    -8-7-6-5-4-3-2-1  0  1  2  3  4  5  6  7  8
   ```

   **A.** $\{-3, 1, 2, -7\}$      **C.** $\{7, -1, 2, -3\}$

   **B.** $\{-3, 7, 1, -2\}$      **D.** $\{-2, 3, 1, 7\}$

   9. _____

# Test, Form 1B   *(continued)*

**For Exercises 10 and 11, use the coordinate plane below.**

10. Which ordered pair represents the point on the coordinate plane?

    **F.** (2, 3)          **H.** (3, 4)

    **G.** (1, 5)          **I.** (0, 2)

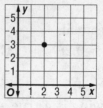

10. _____

11. Which of the following points is located in Quadrant III?

    **A.** (0, 4)          **C.** (−3, −4)

    **B.** (−2, 7)          **D.** (4, −1)

11. _____

12. Which situation does the integer 3 best represent?

    **F.** 3 feet below the ground          **H.** finding $3 in a pocket

    **G.** 3 yards behind the winner          **I.** losing $3

12. _____

**Replace each ● with <, >, or = to make a true statement.**

13. 5 ● −3

    **A.** <          **B.** >          **C.** =          **D.** +

13. _____

14. 11 ● −7

    **F.** <          **G.** >          **H.** =          **I.** +

14. _____

15. Which number is less than −8?

    **A.** 0          **B.** 4          **C.** −7          **D.** −10

15. _____

16. Order −6, 2, 0, and −3 from least to greatest.

    **F.** 2, 0, −6, −3          **H.** −6, −3, 0, 2

    **G.** 0, 2, −3, −6          **I.** 2, 0, −3, −6

16. _____

17. Which quadrant contains the point named by (−1, 3)?

    **A.** Quadrant I          **C.** Quadrant III

    **B.** Quadrant II          **D.** Quadrant IV

17. _____

18. Raymond has $25 to spend and used $17 to buy a new DVD. Which integer best represents the situation of spending $17?

    **F.** 17          **H.** 25

    **G.** −17          **I.** −25

18. _____

19. Courtney used 25 out of her 30 tickets to select a prize at the arcade. What is this fraction written as a decimal?

    **A.** 0.8          **B.** 0.83          **C.** $0.8\overline{3}$          **D.** $0.8\overline{3}$

19. _____

# Test, Form 2A

**Write the letter for the correct answer in the blank at the right of each question.**

1. Which of the following statements is true?

   **A.** $2 < -3$     **B.** $-4 < -5$     **C.** $-4 > -5$     **D.** $-3 > 2$

   1. _____

2. Which set of integers is graphed on the number line?

   $$\overset{\longleftrightarrow}{\underset{-3\,-2\,-1\ \ 0\ \ 1\ \ 2\ \ 3\ \ 4\ \ 5}{\vert\ \vert\ \vert\ \vert\ \vert\ \vert\ \vert\ \vert\ \vert}}$$

   **F.** $\{3, -1, 2\}$    **G.** $\{-3, 1, -2\}$   **H.** $\{-3, -1, 2\}$   **I.** $\{-2, -1, 3\}$

   2. _____

3. Which expression has the greatest value?

   **A.** $-|-16|$     **B.** $|-14|$      **C.** $-|-12|$     **D.** $|10|$

   3. _____

4. What is the value of the expression $|-36| + |7|$?

   **F.** $-43$       **G.** $-29$       **H.** $29$        **I.** $43$

   4. _____

5. Which integer best represents a withdrawal of $85?

   **A.** $85$        **B.** $-85$       **C.** $|85|$       **D.** $|-85|$

   5. _____

6. Write $-\dfrac{5}{11}$ as a decimal.

   **F.** $-0.\overline{4}$      **G.** $-0.\overline{45}$     **H.** $-0.48$     **I.** $-4.5$

   6. _____

7. Order $-3.98$, $3\dfrac{8}{9}$, $-3\dfrac{11}{12}$, and $3.\overline{9}$ from least to greatest.

   **A.** $3.\overline{9}, -3.98, -3\dfrac{11}{12}, 3\dfrac{8}{9}$     **C.** $-3.98, -3\dfrac{11}{12}, 3\dfrac{8}{9}, 3.\overline{9}$

   **B.** $-3.98, 3.\overline{9}, -3\dfrac{11}{12}, 3\dfrac{8}{9}$     **D.** $-3.98, -3\dfrac{8}{9}, -3\dfrac{11}{12}, 3.\overline{9}$

   7. _____

8. What is the opposite of $-132$?

   **F.** $-132$     **G.** $132$      **H.** $0$          **I.** $6$

   8. _____

9. Which integer represents a gain of 7 yards on a play?

   **A.** $+7$       **B.** $+5$       **C.** $-5$       **D.** $-7$

   9. _____

10. Which situation is *not* best described by a negative integer?

    **F.** a withdrawal of $45      **H.** a loss of 12 yards

    **G.** a fine of $15          **I.** a bonus of 10 points

    10. _____

11. Jolene scored 21 goals over the past 18 lacrosse games. What was her average number of goals scored per game expressed as a decimal?

    **A.** $1.15$             **C.** $1.1\overline{6}$

    **B.** $1.16\overline{7}$         **D.** $2.1\overline{6}$

    11. _____

# Test, Form 2A   (continued)

**For Exercises 12 and 13, use the coordinate plane below.**

12. Which of the following correctly identifies
the point for the ordered pair $(-3, 4)$?

   F. point $A$                   H. point $C$
   G. point $B$                   I. point $D$

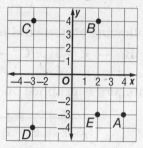

12. _____

13. Which of the following ordered pairs correctly
names point $E$?

   A. $(-2, 3)$                    C. $(-3, 2)$
   B. $(2, -3)$                    D. $(3, -2)$

13. _____

14. Which of the following correctly identifies the quadrant where the
point named by $(9, 3)$ is located?

   F. Quadrant I                   H. Quadrant III
   G. Quadrant II                  I. Quadrant IV

14. _____

**Graph each point on a coordinate plane.**

15. $N(4, 1)$

16. $P(-2, -3)$

17. $Q(3, -2)$

18. $R(-1, 2)$

15–18.

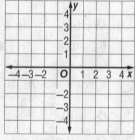

**Replace each ● with <, >, or = to make a true sentence.**

19. $2$ ● $-3$

19. _____

20. $-6$ ● $-4$

20. _____

**For Exercises 21–24, use the coordinate plane below.**

21. Identify the point for the ordered pair $(3, -4)$.

21. _____

22. Write the ordered pair that names point $C$.

22. _____

23. Write the ordered pair that names point $F$.

23. _____

24. Write the ordered pair that represents the
reflection of point $F$ across the $y$-axis.

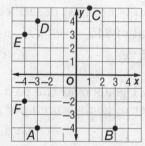

24. _____

# Test, Form 2B

**Write the letter for the correct answer in the blank at the right of each question.**

1. Which of the following statements is true?

   **A.** $4 < -8$     **B.** $-1 < -4$     **C.** $-9 > 0$     **D.** $-2 > -7$

   1. _____

2. Which set of integers is graphed on the number line?

   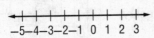

   **F.** $\{1, -1, -5\}$   **G.** $\{-5, -3, 0\}$   **H.** $\{3, -5, 0\}$    **I.** $\{-1, -5, -3\}$

   2. _____

3. Which expression has the greatest value?

   **A.** $-|-13|$     **B.** $|-1|$     **C.** $-|-22|$     **D.** $|20|$

   3. _____

4. What is the value of the expression $|-31| + |-9|$?

   **F.** $-40$     **G.** $-22$     **H.** $29$     **I.** $40$

   4. _____

5. Which integer best represents a deposit of $37?

   **A.** $37$     **B.** $-37$     **C.** $-|37|$     **D.** $-|-37|$

   5. _____

6. Write $-\dfrac{2}{9}$ as a decimal.

   **F.** $-0.\overline{2}$     **G.** $-0.2\overline{3}$     **H.** $-0.22$     **I.** $-2.2$

   6. _____

7. Order $-2.96$, $2\dfrac{1}{9}$, $-2\dfrac{11}{12}$, and $2.\overline{95}$ from least to greatest.

   **A.** $-2.96, 2.\overline{95}, -2\dfrac{11}{12}, 2\dfrac{1}{9}$     **C.** $-2.96, -2\dfrac{11}{12}, 2\dfrac{1}{9}, 2.\overline{95}$

   **B.** $2.\overline{95}, -2.96, -2\dfrac{11}{12}, 2\dfrac{1}{9}$     **D.** $-2.96, -2\dfrac{1}{9}, -2\dfrac{11}{12}, 2.\overline{95}$

   7. _____

8. What is the opposite of 89?

   **F.** $-89$     **G.** $89$     **H.** $0$     **I.** $98$

   8. _____

9. Which integer represents a decrease of five degrees?

   **A.** $+7$     **B.** $+5$     **C.** $-5$     **D.** $-7$

   9. _____

10. Which situation is *not* best described by a negative integer?

    **F.** a height of 75 yards     **H.** a loss of 9 pounds

    **G.** a decrease of 4 points     **I.** 3 degrees below zero

    10. _____

11. Which of the following correctly identifies the quadrant where the point named by (2, 2) is located?

    **A.** Quadrant I     **C.** Quadrant III

    **B.** Quadrant II     **D.** Quadrant IV

    11. _____

# Test, Form 2B   (continued)

**For Exercises 12 and 13, use the coordinate plane below.**

12. Which of the following correctly identifies
the point for the ordered pair (4, −3)?

    **F.** point *A*          **H.** point *C*
    **G.** point *B*          **I.** point *D*

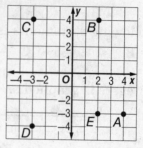

12. _____

13. Which of the following ordered pairs correctly
names point *D*?

    **A.** (−3, 4)          **C.** (−3, −4)
    **B.** (3, −4)          **D.** (3, 4)

13. _____

14. Which of the following correctly identifies the quadrant where the
point named by (−3, 7) is located?

    **F.** Quadrant I          **H.** Quadrant III
    **G.** Quadrant II          **I.** Quadrant IV

14. _____

**Graph each point on a coordinate plane.**

15. *M*(0, 3)

16. *R*(−3, 3)

17. *S*(−2, −4)

18. *T*(4, −1)

**15–18.**

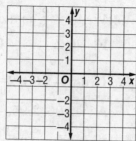

**Replace each ● with <, >, or = to make a true sentence.**

19. 0 ● −1

20. −7 ● −9

19. _____

20. _____

**For Exercises 21–24, use the coordinate plane below.**

21. Identify the point for the ordered pair (−4, 3).

22. Write the ordered pair that names point *A*.

23. Write the ordered pair that names point *D*.

24. Write the ordered pair that represents the
reflection of point *D* across the *y*-axis.

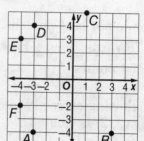

21. _____

22. _____

23. _____

24. _____

# Test, Form 3A

**Write an integer to describe each situation.**

1. Jane lost a twenty dollar bill.

1. _____

2. The balloon expands three centimeters each second.

2. _____

3. Graph the set of integers −3, −1, and 2 on the number line.

3.

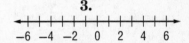

**Evaluate.**

4. $|12| + |-63|$

4. _____

5. $|-8| + |-2|$

5. _____

6. $|-4| - |-1|$

6. _____

7. $|3| - |-2|$

7. _____

8. $|-19| + |4|$

8. _____

9. The locations of three fish relative to the water's surface are −18 feet, −31 feet, and −26 feet. Which distance has the least absolute value?

9. _____

10. Graph the set {−4, −3, 1} on the number line.

10.

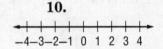

**Replace each ● with <, >, or = to make a true sentence.**

11. −8 ● −9

11. _____

12. 1 ● −4

12. _____

13. −3 ● 5

13. _____

14. $|-7|$ ● −7

14. _____

15. $|-8|$ ● $|8|$

15. _____

16. Order −7, 4, −2, 0, and −5 from least to greatest.

16. _____

# Test, Form 3A   (continued)

17. Write $-\dfrac{7}{11}$ as a decimal.

17. _____

18. Write $-\dfrac{18}{25}$ as a decimal.

18. _____

**Order the numbers from least to greatest.**

19. $7\dfrac{6}{11}$, $-7.6$, $7.\overline{34}$, $-7\dfrac{4}{5}$

19. _____

20. $-10\dfrac{1}{5}$, $10.19$, $-10\dfrac{2}{11}$, $-10.\overline{20}$

20. _____

21. Graph and label the points $A(4, 2)$, $B(-4, 1)$, and $C(-3, -2)$ on the coordinate plane.

21.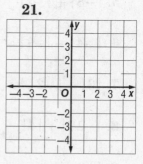

**For Exercises 22–24, use the coordinate plane below.**

22. Identify the point for the ordered pair $(-4, 2)$.

22. _____

23. Write the ordered pair that names point $B$.

23. _____

24. Write the ordered pair that names point $D$.

24. _____

25. In which quadrant is the point at $(-4, 5)$ located?

25. _____

**For Exercises 26–28, use the coordinate plane below that represents the location of a swimming pool.**

26. An oak tree is located at the reflection of point $B$ across the $x$-axis. What ordered pair describes the location of the oak tree?

26. _____

27. A sprinkler is located at the reflection of point $C$ across the $x$-axis. What ordered pair describes the location of the sprinkler?

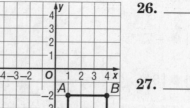

27. _____

28. A fire pit is located at the reflection of point $D$ about the $y$-axis. What ordered pair describes the location of the fire pit?

28. _____

# Test, Form 3B

**Write an integer to describe each situation.**

1. Matt found a five dollar bill.

2. The tub drains 4 inches each minute.

3. Graph the set of integers −5, −1, and 2 on the number line.

**Evaluate.**

4. |14| + |−31|

5. |−2| + |−9|

6. |−6| − |−2|

7. |8| − |−5|

8. |−22| + |6|

9. The locations of three fish relative to the water's surface are −13 feet, −33 feet, and −22 feet. Which distance has the least absolute value?

10. Graph the set {−2, −1, 3} on the number line.

**Replace each ● with <, >, or = to make a true sentence.**

11. −2 ● −7

12. 5 ● −1

13. −7 ● 11

14. |−6| ● 6

15. |−4| ● −|4|

16. Order −6, 3, −1, 0, and −4 from least to greatest.

1. _____

2. _____

3.
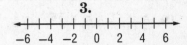

4. _____

5. _____

6. _____

7. _____

8. _____

9. _____

10.
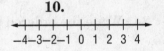

11. _____

12. _____

13. _____

14. _____

15. _____

16. _____

# Test, Form 3B   (continued)

17. Write $-\dfrac{7}{15}$ as a decimal.

17. _____

18. Write $-\dfrac{13}{20}$ as a decimal.

18. _____

**Order the numbers from least to greatest.**

19. $5\dfrac{5}{11}$, $-5.5$, $5.\overline{33}$, $-5\dfrac{4}{5}$

19. _____

20. $-8\dfrac{1}{5}$, $8.23$, $-8\dfrac{2}{11}$, $-8.\overline{21}$

20. _____

21. Graph and label the points $A(-1, 4)$, $B(2, -3)$, and $C(2, 3)$ on the coordinate plane.

21.

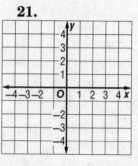

**For Exercises 22–24, use the coordinate plane below.**

22. Identify the point for the ordered pair $(4, -2)$.

22. _____

23. Write the ordered pair that names point $E$.

23. _____

24. Write the ordered pair that names point $F$.

24. _____

25. In which quadrant is the point at $(-2, -3)$ located?

25. _____

**For Exercises 26–28, use the coordinate plane below that represents the location of a library.**

26. A post office is located at the reflection of point $A$ across the $x$-axis. What ordered pair describes the location of the post office?

26. _____

27. A candy store is located at the reflection of point $D$ across the $x$-axis. What ordered pair describes the location of the candy store?

27. _____

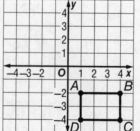

28. A school is located at the reflection of point $B$ about the $y$-axis. What ordered pair describes the location of the school?

28. _____

# Are You Ready?

## Review

> The side lengths of a square are all the same length. To find the **area** of a square you multiply the lengths of the two sides together.

### Example 1

**Find the area of a square shown at the right.**

$A = s \times s$      Area of a square
$A = 12 \times 12$      Replace s with 12.
$A = 144$      Multiply.

The area of the square is 144 square centimeters.

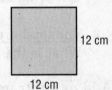

12 cm

12 cm

### Example 2

**Find the area of a square with a side length of 6 inches.**

$A = s \times s$      Area of a square
$A = 6 \times 6$      Replace s with 6.
$A = 36$      Multiply.

The area of the square is 36 square inches.

**Find the area of each square.**

1.

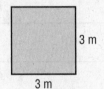

3 m

3 m

1. _____

2.

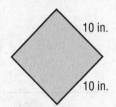

10 in.

10 in.

2. _____

3.

4 ft

4 ft

3. _____

4. Find the area of a square with a side length of 11 feet.

4. _____

5. Find the area of a square with a side length of 7 meters.

5. _____

6. Find the area of a square with a side length of 13 centimeters.

6. _____

# Are You Ready?

## Practice

1. Find the area of a square with a side length of 9 feet.

1. _____

2. Find the area of a square with a side length of 11 centimeters.

2. _____

3. The total area of a square is 25 square meters. What is the length of the side?

3. _____

**Add or subtract. Write in simplest form.**

4. $\dfrac{3}{4} + \dfrac{1}{4}$

4. _____

5. $\dfrac{7}{8} - \dfrac{1}{8}$

5. _____

6. $1 - \dfrac{5}{9}$

6. _____

7. $\dfrac{5}{7} + \dfrac{5}{7}$

7. _____

8. $\dfrac{2}{9} - \dfrac{1}{18}$

8. _____

9. $7\dfrac{2}{3} + \dfrac{1}{4}$

9. _____

10. $7\dfrac{1}{6} - 2\dfrac{1}{5}$

10. _____

11. **SAVINGS** Angie has saved $\dfrac{3}{8}$ of the amount she needs to buy a bike. Clive has saved $\dfrac{7}{10}$ of the amount he needs to buy a skateboard. What is the difference between Clive's and Angie's fraction of savings?

11. _____

12. **TRANSPORTATION** Mr. Ping took a taxi $4\dfrac{1}{2}$ miles to the train station, a train $32\dfrac{7}{8}$ miles to an appointment, and then a bus $15\dfrac{3}{4}$ miles to his business. How far did he travel in all?

12. _____

# Are You Ready?

## Apply

| | |
|---|---|
| **1. GEOMETRY** Find the area of a square with a length of 6 inches. | **2. TABLECLOTHS** The area of a square tablecloth is 16 square feet. What is the length of the side of the tablecloth? |
| **3. BAKING** Katiana needs to bake a dozen muffins of a dozen different flavors for a school bake sale. How many muffins does Katiana need to bake? | **4. WATER** Shasta poured water into three one-gallon water jugs to take to a race. She filled the first jug $\frac{3}{4}$ full. She filled each of the second and third jugs $\frac{7}{8}$ full. How much water did Shasta take to the race? |
| **5. WALLPAPER** Yoki is putting up new wallpaper in her room. She wants to add a border along the ceiling. If her room is a rectangle with sides of $7\frac{1}{2}$ feet and $9\frac{3}{4}$ feet, how long of a border will she need? | **6. RESTAURANT** A restaurant sells pies by the slice. At the end of the night they have $\frac{1}{2}$ of a cherry pie, $\frac{2}{3}$ of an apple pie, and $\frac{1}{6}$ of a banana cream pie. How much total pie is left? |

# Diagnostic Test

**1.** Find the area of the square shown below.

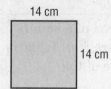

14 cm

14 cm

1. _____

**2.** Find the area of a square with a side length of 15 inches.

2. _____

**3. GEOMETRY** The area of a square is 4 square miles. What is length of the side of the square?

3. _____

**Add or subtract. Write in simplest form.**

**4.** $\frac{8}{16} + \frac{2}{16}$

4. _____

**5.** $\frac{11}{20} - \frac{3}{20}$

5. _____

**6.** $\frac{7}{9} - \frac{1}{3}$

6. _____

**7.** $3\frac{2}{7} + \frac{1}{4}$

7. _____

**8.** $5\frac{1}{3} - 1\frac{3}{10}$

8. _____

**9.** $10\frac{1}{2} + 1\frac{1}{4}$

9. _____

**10. FUNDRAISING** Domanick made $\frac{1}{4}$ of the total sales and Billy made $\frac{7}{40}$ of the total sales. What is the difference between Domanick's and Billy's fraction of sales?

10. _____

**11. GEOMETRY** To find the perimeter of a triangle, find the sum of the measures of the three sides. Find the perimeter of the triangle shown.

11. _____

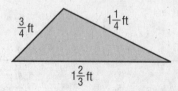

$\frac{3}{4}$ ft     $1\frac{1}{4}$ ft

$1\frac{2}{3}$ ft

# Pretest

**Find the value of each expression.**

1. $8 + 9 \times 3 - 4$

   1. _____

2. $16 - 3^2$

   2. _____

**Evaluate the expression if $a = 4$ and $b = 3$**

3. $a + 8 - b$

   3. _____

4. $5a$

   4. _____

**Define a variable then write each phrase as an algebraic expression.**

5. 9 years older than Odile

   5. _____

6. one half of the total cost

   6. _____

7. three times the number of berries

   7. _____

**Determine whether the two expressions are equivalent. If so, tell what property is applied. If not, explain why.**

8. $19 \times 6$ and $6 \times 19$

   8. _____

9. $14$ and $14 + 0$

   9. _____

**Use the Distributive Property to rewrite each algebraic expression.**

10. $6(d + 7)$

    10. _____

11. $2(a + 10)$

    11. _____

12. $5(y - 5)$

    12. _____

# Chapter Quiz

**Write each product using an exponent.**

**1.** $9 \times 9 \times 9 \times 9$

**2.** $17 \times 17 \times 17$

**3.** $4 \times 4 \times 4 \times 4 \times 4$

**4.** $6 \times 6$

**Write each power as a product of the same factor. Then find the value.**

**5.** $5^3$

**6.** $4^6$

**7.** $7^4$

**8.** $11^2$

**Find the value of each expression.**

**9.** $3 + 12 - 9$        **10.** $(7 + 4) \times 3 - 2$

**11.** $2^4 - 6 \div 2$        **12.** $5 + 4 \times (2 + 7)$

**Evaluate each expression if $m = 3$ and $n = 7$.**

**13.** $n + 9$        **14.** $4m - 5$

**15.** $n^2 - 2m$        **16.** $3mn$

**Write each phrase as an algebraic expression.**

**17.** four times a number        **18.** a number divided by 14

1. _____

2. _____

3. _____

4. _____

5. _____

6. _____

7. _____

8. _____

9. _____

10. _____

11. _____

12. _____

13. _____

14. _____

15. _____

16. _____

17. _____

18. _____

# Vocabulary Test

| | | |
|---|---|---|
| algebraic expression | Distributive Property | numerical expressions |
| Associative Property | equivalent expressions | order of operations |
| base | evaluate | perfect square |
| coefficient | exponent | powers |
| Commutative Property | factor the expression | properties |
| constant | Identity Property | term |
| define the variable | like terms | variable |

## Choose from the terms above to complete each sentence.

1. _____ are combinations of numbers and operations.

1. _____

2. The _____ tells you which operation to perform first.

2. _____

3. A _____ is a symbol, usually a letter, used to represent a number.

3. _____

4. _____ are statements that are true for any number.

4. _____

5. $2(c + 7) = 2c + 14$ is an example of the _____.

5. _____

6. The expression $x + 16 \div 2$ is an example of an _____.

6. _____

7. $6 + 19 = 19 + 6$ is an example of the _____ Property.

7. _____

8. $1 + (3 + 8) = (1 + 3) + 8$ is an example of the _____ Property.

8. _____

## Define each term in your own words.

9. base

9. _____

10. exponent

10. _____

# Standardized Test Practice

**Read each question. Then fill in the correct answer on the answer sheet provided by your teacher or on a sheet of paper.**

1. The perimeter of a rectangle can be found using the expression $2\ell + 2w$, where $\ell$ represents length and $w$ represents width. Find the perimeter of the front of a new building whose design is shown below.

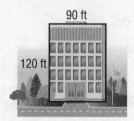
90 ft
120 ft

  **A.** 180 feet       **C.** 240 feet

  **B.** 210 feet       **D.** 420 feet

2. Ora bought the variety pack of granola bars shown. The box contains 24 granola bars. How much does one granola bar cost?

$7.20
Granola Bars
Variety Pack
24 bars

  **F.** $0.29

  **G.** $0.30

  **H.** $1.35

  **I.** $3.33

3. The cost of renting roller blades is $4 plus $3.50 for each hour that the roller blades are rented. Which expression can be used to find the cost in dollars of renting roller blades for $h$ hours?

  **A.** $4h + 3.5$

  **B.** $3.5 - 4h$

  **C.** $3.5(h + 4)$

  **D.** $3.5h + 4$

4. ▰▱ **GRIDDED RESPONSE** The Manny family is installing a large patio in their backyard. Find the area of the patio in square yards.

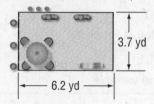
3.7 yd
6.2 yd

5. Which of the following illustrates the Distributive Property?

  **F.** $3(2x + 4) = 5x + 4$

  **G.** $3(2x + 4) = 5x + 7$

  **H.** $3(2x + 4) = 6x + 4$

  **I.** $3(2x + 4) = 6x + 12$

6. **THINK SOLVE EXPLAIN** **SHORT RESPONSE** What is the value of $45 \div (7 + 2) - 1$?

7. The table shows the constant speed that each driver is driving.

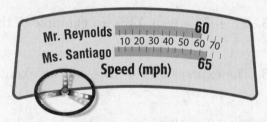
Mr. Reynolds   60
10 20 30 40 50 60 70
Ms. Santiago   65
Speed (mph)

Using the Distributive Property, how many more miles will Ms. Santiago drive in 3 hours than Mr. Reynolds?

  **A.** 5 miles

  **B.** 15 miles

  **C.** 180 miles

  **D.** 195 miles

8. Each figure below is divided into sections of equal size. Which figure has 87.5% of its total area shaded?

F.

H.

G.

I.

9. Which expression is NOT an example of the Commutative Property?

A. $b - t = t - b$     C. $f + a = a + f$

B. $rt = tr$          D. $5 \cdot d = d \cdot 5$

10. At a state fair, each person pays $8 for admission plus $2 for each ride. While at the fair, Elizabeth goes on 6 rides. Which expression can be used to find the total amount Elizabeth spends?

F. $\$8 + 6 \times \$2$     H. $(\$8 + \$2) + 6$

G. $(\$8 + \$2) \times 6$     I. $\$8 \times 6 \times \$2$

11. ✎ **GRIDDED RESPONSE** Norene is buying a coat that is on sale. Write 30% as a decimal.

12. Which expression is equivalent to $5 + 4^2 \times 2$?

A. $21 \times 2$     C. $9^2 \times 2$

B. $5 + 32$          D. $5 + 8^2$

13. ✎ **GRIDDED RESPONSE** A garden has five light poles in the positions shown.

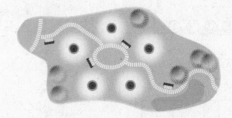

A party planner wants to connect each light pole directly to each of the other poles with a string of lights for an outdoor party. The expression $\frac{n(n-1)}{2}$, where $n$ represents the number of poles, can be used to determine how many strings are needed. How many strings are needed to connect the five poles?

14. **THINK SOLVE EXPLAIN** **EXTENDED RESPONSE** Cary earns $5 per hour raking leaves for her next-door neighbor. She owes her mom $6.

**Part A** Suppose Cary raked leaves for $h$ hours and paid her mother the $6 she owed her. Write an expression to represent the amount of money Cary will have left.

**Part B** Use your expression from *Part A* to find how much money she will have left if she rakes leaves for 3 hours and pays her mother. Explain how you solved.

# Student Recording Sheet

*Use this recording sheet with the Standardized Test Practice pages.*

**Fill in the correct answer. For gridded-response questions, write your answers in the boxes on the answer grid and fill in the bubbles to match your answers.**

1. Ⓐ Ⓑ Ⓒ Ⓓ

2. Ⓕ Ⓖ Ⓗ Ⓘ

3. Ⓐ Ⓑ Ⓒ Ⓓ

4.

5. Ⓕ Ⓖ Ⓗ Ⓘ

6. _____

7. Ⓐ Ⓑ Ⓒ Ⓓ

8. Ⓕ Ⓖ Ⓗ Ⓘ

9. Ⓐ Ⓑ Ⓒ Ⓓ

10. Ⓕ Ⓖ Ⓗ Ⓘ

11.

12. Ⓐ Ⓑ Ⓒ Ⓓ

13.

## Extended Response

**Record your answers for Exercise 14 on the back of this paper.**

# Extended-Response Test

**Demonstrate your knowledge by giving a clear, concise solution to each problem. Be sure to include all relevant drawings and justify your answers. You may show your solution in more than one way or investigate beyond the requirements of the problem. If necessary, record your answer on another piece of paper.**

1. Write the order of operations in your own words.

2. Every winter, students at Camden Middle School go on a class ski trip. Fifteen students always sign up first. For every inch of snow that falls, an additional 25 students sign up.

   a. Write an expression to show the total number of students going on the trip, using only a variable to represent the additional students.

   b. Now write a different expression to show the total number of students going on the trip, using an expression consisting of a variable and a number to represent the additional students.

   c. Are the expressions equivalent? Explain.

3. Evaluate $7p + 6(p \div q)^2 - 2q$ if $p = 6$ and $q = 3$. Show your work and give an explanation for each step.

4. Use the properties you learned to show why $(a + b) + c$ is equivalent to $b + (a + c)$.

# Extended-Response Rubric

| Score | Description |
|---|---|
| 4 | A score of four is a response in which the student demonstrates a thorough understanding of the mathematics concepts and/or procedures embodied in the task. The student has responded correctly to the task, used mathematically sound procedures, and provided clear and complete explanations and interpretations. The response may contain minor flaws that do not detract from the demonstration of a thorough understanding. |
| 3 | A score of three is a response in which the student demonstrates an understanding of the mathematics concepts and/or procedures embodied in the task. The student's response to the task is essentially correct with the mathematical procedures used and the explanations and interpretations provided demonstrating an essential but less than thorough understanding. The response may contain minor flaws that reflect inattentive execution of mathematical procedures or indications of some misunderstanding of the underlying mathematics concepts and/or procedures. |
| 2 | A score of two indicates that the student has demonstrated only a partial understanding of the mathematics concepts and/or procedures embodied in the task. Although the student may have used the correct approach to obtaining a solution or may have provided a correct solution, the student's work lacks an essential understanding of the underlying mathematical concepts. The response contains errors related to misunderstanding important aspects of the task, misuse of mathematical procedures, or faulty interpretations of results. |
| 1 | A score of one indicates that the student has demonstrated a very limited understanding of the mathematics concepts and/or procedures embodied in the task. The student's response is incomplete and exhibits many flaws. Although the student's response has addressed some of the conditions of the task, the student reached an inadequate conclusion and/or provided reasoning that was faulty or incomplete. The response exhibits many flaws or may be incomplete. |
| 0 | A score of zero indicates that the student has provided no response at all, or a completely incorrect or uninterpretable response, or demonstrated insufficient understanding of the mathematics concepts and/or procedures embodied in the task. For example, a student may provide some work that is mathematically correct, but the work does not demonstrate even a rudimentary understanding of the primary focus of the task. |

# Test, Form 1A

**Write the letter for the correct answer in the blank at the right of each question.**

1. The volume of a certain cube can be found using the expression $5^3$. What is $5^3$ written as a product of the same factor?

   **A.** $5 \times 3$            **C.** $3 \times 5$

   **B.** $3 \times 3 \times 3 \times 3 \times 3$      **D.** $5 \times 5 \times 5$            1. _____

2. What is $8 \times 8 \times 8 \times 8$ written using an exponent?

   **F.** $8^4$       **G.** $4^8$       **H.** $8 \times 4$       **I.** $4{,}096$            2. _____

**What is the value of each expression?**

3. $5^2 + 7$

   **A.** 12        **B.** 32        **C.** 33        **D.** 42            3. _____

4. $21 - 3^2 + 2$

   **F.** 14        **G.** 16        **H.** 20        **I.** 24            4. _____

5. $58 - 2 \times 3 + 1$

   **A.** 50        **B.** 53        **C.** 169        **D.** 224            5. _____

6. $4 \times 3 + 9 \times 8$

   **F.** 59        **G.** 84        **H.** 168        **I.** 384            6. _____

7. What is the value of $cd$ if $c = 9$ and $d = 8$?

   **A.** 98        **B.** 89        **C.** 72        **D.** 17            7. _____

8. What is the value of $2 + 3n$ if $n = \frac{1}{2}$?

   **F.** $1\frac{1}{2}$       **G.** $3\frac{1}{2}$       **H.** $5\frac{1}{2}$       **I.** 8            8. _____

9. What is the value of $s + t - u$ if $s = 12$, $t = 8$, and $u = 20$?

   **A.** 0        **B.** 10        **C.** 15        **D.** 18            9. _____

**Which is the correct algebraic expression for each phrase?**

10. 10 dollars less than Caitlin

    **F.** $c + 10$      **G.** $c - 10$      **H.** $10 - c$      **I.** $10c$            10. _____

11. 13 times the cost of one ticket

    **A.** $t \div 13$      **B.** $t - 13$      **C.** $13 + t$      **D.** $13t$            11. _____

12. twelve inches longer than the width

    **F.** $12w$      **G.** $12 - w$      **H.** $w + 12$      **I.** $12 \div w$            12. _____

13. Which property is illustrated by the statement $3 + 0 = 3$?
    **A.** Associative        **C.** Distributive
    **B.** Commutative        **D.** Identity                    13. _____

14. Which property is illustrated by the statement $6 \cdot 4 = 4 \cdot 6$?
    **F.** Associative        **H.** Distributive
    **G.** Commutative        **I.** Identity                    14. _____

15. Which of the following is equivalent to $2 \cdot (4 \cdot 3)$?
    **A.** $2 + (4 + 3)$ **B.** $2 \cdot (6 \cdot 4)$ **C.** $(2 \cdot 4) \cdot 3$ **D.** 8    15. _____

16. Which of the following is the factored form of the expression $18 + 12$?
    **F.** $2(9 + 6)$  **G.** $3(6 + 4)$  **H.** $6(3 + 2)$  **I.** $9(2 + 3)$    16. _____

17. Which shows how to find $5 \times 83$ mentally by using the Distributive Property?
    **A.** $3(5 + 80)$  **B.** $5(80) + 3$  **C.** $3(80) + 5(3)$ **D.** $5(80) + 5(3)$    17. _____

**Which expression results from using the Distributive Property?**

18. $6(x + 4)$
    **F.** $6x + 10$  **G.** $6x + 4$  **H.** $24x$  **I.** $6x + 24$    18. _____

19. $2(5 + r)$
    **A.** $7 + r$  **B.** $10 + 2r$  **C.** $12r$  **D.** $7 + 2r$    19. _____

20. $11(n + 3)$
    **F.** $14n$  **G.** $n + 33$  **H.** $33n$  **I.** $11n + 33$    20. _____

**What is the simplified form of each expression?**

21. $2x + 5x + 4x$
    **A.** $11 + 3x$  **B.** $7x$  **C.** $11x$  **D.** $7x + 4x$    21. _____

22. $5(4x)$
    **F.** $9x$  **G.** $5(4) + 5(x)$ **H.** $9 + x$  **I.** $20x$    22. _____

23. $7(2x + 6y)$
    **A.** $14x + 42y$ **B.** $56x$  **C.** $56xy$  **D.** $14x + 42$    23. _____

**What is the factored form of each expression?**

24. $20x + 35y$
    **F.** $4x + 7y$  **G.** $5xy(4 + 7)$ **H.** $5(4x + 7y)$  **I.** $(20 + 35) \cdot (x + y)$    24. _____

25. $24x + 64y$
    **A.** $4(6x + 16y)$ **B.** $8(3x + 8y)$ **C.** $8xy(3 + 8)$ **D.** $3x + 8y$    25. _____

# Test, Form 1B

**Write the letter for the correct answer in the blank at the right of each question.**

1. The volume of a certain cube can be found using the expression $8^3$.
   What is $8^3$ written as a product of the same factor?
   **A.** $8 \times 8 \times 8$       **C.** $3 \times 3 \times 3 \times 3 \times 3 \times 3 \times 3 \times 3$
   **B.** $3 \times 8$       **D.** $8 \times 3$

   1. _____

2. What is $4 \times 4 \times 4 \times 4 \times 4$ written using an exponent?
   **F.** 1,024       **G.** $4^5$       **H.** $4 \times 5$       **I.** $5^4$

   2. _____

**What is the value of each expression?**

3. $6^2 + 3$
   **A.** 15       **B.** 16       **C.** 39       **D.** 42

   3. _____

4. $31 + 5^2 - 6$
   **F.** 40       **G.** 50       **H.** 52       **I.** 62

   4. _____

5. $46 - 2 \times 4 + 7$
   **A.** 38       **B.** 45       **C.** 183       **D.** 484

   5. _____

6. $5 \times 4 + 8 \times 7$
   **F.** 168       **G.** 84       **H.** 76       **I.** 59

   6. _____

7. What is the value of $cd$ if $c = 7$ and $d = 6$?
   **A.** 13       **B.** 42       **C.** 67       **D.** 76

   7. _____

8. What is the value of $5 + 2n$ if $n = \frac{3}{4}$?
   **F.** $1\frac{1}{2}$       **G.** $5\frac{3}{4}$       **H.** $6\frac{1}{2}$       **I.** $7\frac{3}{4}$

   8. _____

9. What is the value of $s - t + u$ if $s = 11$, $t = 9$, and $u = 10$?
   **A.** 19       **B.** 15       **C.** 12       **D.** 8

   9. _____

**Which is the correct algebraic expression for each phrase?**

10. 14 more pickles than the first jar
    **F.** $p + 14$       **G.** $14 - p$       **H.** $14p$       **I.** $14 \div p$

    10. _____

11. 7 inches shorter than Sue
    **A.** $s - 7$       **B.** $7 + s$       **C.** $7 - s$       **D.** $s \div 7$

    11. _____

12. 10 times the number of marbles Yolanda has
    **F.** $m \div 10$       **G.** $m - 10$       **H.** $10 + m$       **I.** $10m$

    12. _____

# Test, Form 1B    (continued)

**13.** Which property is illustrated by the statement $8 + 2 = 2 + 8$?

   **A.** Associative          **C.** Distributive

   **B.** Commutative       **D.** Identity                 **13.** _____

**14.** Which property is illustrated by the statement $3 \cdot 1 = 3$?

   **F.** Associative           **H.** Distributive

   **G.** Commutative        **I.** Identity                  **14.** _____

**15.** Which of the following is equivalent to $5 \cdot (7 \cdot 4)$?

   **A.** 35        **B.** $(5 \cdot 7) \cdot 4$    **C.** $5 \cdot (4 \cdot 4)$    **D.** $5 + (7 + 4)$    **15.** _____

**16.** Which of the following is the factored form of the expression $20 + 16$?

   **F.** $2(10 + 8)$    **G.** $4(5 + 4)$    **H.** $4(2 + 3)$     **I.** $(5 + 4)$       **16.** _____

**17.** Which shows how to find $5 \times 72$ mentally by using the Distributive Property?

   **A.** $5(7) + 5(2)$   **B.** $5(70) + 5(2)$  **C.** $2(5 + 70)$    **D.** $5(70) + 2$     **17.** _____

## Which expression results from using the Distributive Property?

**18.** $7(x + 3)$

   **F.** $7x + 21$     **G.** $7x + 3$      **H.** $21x$           **I.** $x + 21$       **18.** _____

**19.** $3(6 + r)$

   **A.** $6 + r$       **B.** $6 + 3r$      **C.** $18r$          **D.** $18 + 3r$      **19.** _____

**20.** $12(n + 2)$

   **F.** $12n$          **G.** $n + 24$     **H.** $12n + 24$    **I.** $24n$         **20.** _____

## What is the simplified form of each expression?

**21.** $3x + 7x + 2x$

   **A.** $12x$        **B.** $10x + 2x$   **C.** $12 + 5x$    **D.** $10x$        **21.** _____

**22.** $2(8x)$

   **F.** $16x$         **G.** $2(8) + 2(x)$   **H.** $10 + x$    **I.** $10x$         **22.** _____

**23.** $4(3x + 2y)$

   **A.** $20x$        **B.** $12x + 8y$   **C.** $20xy$     **D.** $12x + 8$     **23.** _____

## What is the factored form of each expression?

**24.** $14x + 21y$

   **F.** $7(2x + 3y)$   **G.** $7xy(2 + 3)$   **H.** $2x + 3y$      **I.** $(14 + 21) \cdot (x + y)$    **24.** _____

**25.** $32x + 48y$

   **A.** $16(2x + 3y)$  **B.** $8(4x + 6y)$    **C.** $16xy(2 + 3)$  **D.** $2x + 3y$     **25.** _____

# Test, Form 2A

**Write the letter for the correct answer in the blank at the right of each question.**

1. What is $6^4$ written as a product of the same factor?
   - **A.** $4 \times 4 \times 4 \times 4 \times 4 \times 4$
   - **C.** $6 \times 6 \times 6 \times 6$
   - **B.** $6 \times 4$
   - **D.** $6 + 6 + 6 + 6$

   1. _____

2. What is $7 \times 7 \times 7$ written using an exponent?
   - **F.** $3^7$
   - **G.** 343
   - **H.** $7 \times 3$
   - **I.** $7^3$

   2. _____

**What is the value of each expression?**

3. $18 + 2 \times 3$
   - **A.** 60
   - **B.** 56
   - **C.** 24
   - **D.** 23

   3. _____

4. $2^4 - 3 + 2 \times 5$
   - **F.** 75
   - **G.** 35
   - **H.** 23
   - **I.** 15

   4. _____

5. $11 \times 12 + 3^3 \div 3$
   - **A.** 141
   - **B.** 135
   - **C.** 53
   - **D.** 47

   5. _____

6. $(2^5 - 4) - 3 \times (2 + 1)$
   - **F.** 7
   - **G.** 19
   - **H.** 51
   - **I.** 75

   6. _____

7. A theater charges $9.50 for adults and $6 for children. Which expression shows the total cost of buying 3 adult tickets and 2 children's tickets?
   - **A.** $9.50 \times \$6 + 3 \times 2$
   - **C.** $9.50 + \$6 \times 2 \times 3$
   - **B.** $9.50 \times 2 + \$6 \times 3$
   - **D.** $9.50 \times 3 + \$6 \times 2$

   7. _____

8. What is the value of $x \div y - z$ if $x = 32$, $y = 4$, and $z = 2\frac{3}{4}$?
   - **F.** 6
   - **G.** $5\frac{1}{4}$
   - **H.** $10\frac{3}{4}$
   - **I.** 22

   8. _____

9. Jamie has saved $42 of her allowance money to buy books. If she buys 5 books at $d$ dollars per book, she will have $42 - 5d$ of her allowance left. How much does she have left if the books cost $3.50 each?
   - **A.** $17.50
   - **B.** $24.50
   - **C.** $33.50
   - **D.** $37

   9. _____

**Which is the correct algebraic expression for each phrase?**

10. 15 miles less than Devra biked
    - **F.** $d + 15$
    - **G.** $d - 15$
    - **H.** $15 - d$
    - **I.** $15 + d$

    10. _____

11. one half the number of quarters Diane has
    - **A.** $2d$
    - **B.** $d - 2$
    - **C.** $2 \div d$
    - **D.** $d \div 2$

    11. _____

12. 2.5 times as many pages
    - **F.** $2.5p$
    - **G.** $p + 2.5$
    - **H.** $2.5 \div p$
    - **I.** $p - 2.5$

    12. _____

# Test, Form 2A    (continued)

**13.** Sergei caught 4 striped fish. Each one weighed the same number of pounds. Then he caught another fish that weighed 8 pounds. Which expression represents the total weight of the fish that Sergei caught?

   **A.** $4 + 8 + p$    **B.** $4 + 8p$        **C.** $4p + 8$        **D.** $4p + 8p$

13. _____

**14.** Which property is illustrated by the statement $13 \times 12 = 12 \times 13$?

   **F.** Associative  **G.** Identity    **H.** Distributive  **I.** Commutative

14. _____

**15.** Which property is illustrated by the statement $(3 \times 6) \times 4 = 3 \times (6 \times 4)$?

   **A.** Associative  **B.** Identity      **C.** Distributive  **D.** Commutative

15. _____

**16.** Which shows how to find $7 \times 210$ mentally by using the Distributive Property?

   **F.** $7(200) + 10$              **G.** $7(200) + 7(10)$

   **H.** $7(21) + 7(10)$            **I.** $10(7 + 200)$

16. _____

**17.** Six students each ordered a bagel for $1.20 and a carton of milk for $0.80. Which expression cannot be used to find the total cost of the six breakfasts?

   **A.** $6(\$1.20) + 6(\$0.80)$          **C.** $6(\$1.80)$

   **B.** $6(\$1.20 + \$0.80)$            **D.** $6(\$2.00)$

17. _____

**Use the Distributive Property to rewrite each expression.**

**18.** $3(x + 12)$

18. _____

**19.** $5(8 + r)$

19. _____

**20.** Find the value of $14 \times 3 + 12 \div 4$.

20. _____

**21.** Evaluate $hj - h$ if $h = 4$ and $j = 6$.

21. _____

**Determine whether the two expressions are equivalent. If so, tell what property is applied. If not, explain why.**

**22.** $8 \cdot (4 \cdot 3)$ and $(8 \cdot 4) \cdot 3$

22. _____

**23.** $17 + 12$ and $12 + 17$

23. _____

**Simplify each expression.**

**24.** $3x + 6x + 5x$

24. _____

**25.** $7(2x)$

25. _____

**26.** $4(8x + 3y)$

26. _____

**27.** Six friends went to a baseball game. The price of admission per person was $x. Three of the friends bought a hot dog for $3. Write and simplify an expression that represents the total cost.

27. _____

# Test, Form 2B

**Write the letter for the correct answer in the blank at the right of each question.**

**1.** What is $3^6$ written as a product of the same factor?

    **A.** $3 \times 3 \times 3 \times 3 \times 3 \times 3$     **C.** $6 \times 6 \times 6$

    **B.** $3 \times 6$     **D.** $3 + 3 + 3 + 3$

    **1.** _____

**2.** What is $2 \times 2 \times 2$ written using an exponent?

    **F.** $2^3$     **G.** $8$     **H.** $2 \times 3$     **I.** $3^2$

    **2.** _____

**What is the value of each expression?**

**3.** $20 - 2^3 \div 4$

    **A.** $3$     **B.** $18$     **C.** $22$     **D.** $88$

    **3.** _____

**4.** $28 + 6 \times 4 - 2$

    **F.** $44$     **G.** $50$     **H.** $68$     **I.** $134$

    **4.** _____

**5.** $3^4 \div (9 + 3)$

    **A.** $1$     **B.** $2.25$     **C.** $6.75$     **D.** $12$

    **5.** _____

**6.** $42 \times 3 - (2^4 - 1) \div 5$

    **F.** $0$     **G.** $22.2$     **H.** $123$     **I.** $124.6$

    **6.** _____

**7.** A zoo charges \$11.50 for adults and \$8 for children. Which expression shows the total cost of buying 2 adult tickets and 4 children's tickets?

    **A.** \$11.50 $\times$ 2 + \$8 $\times$ 4     **C.** \$11.50 $\times$ \$8 + 2 $\times$ 4

    **B.** \$11.50 $\times$ 4 + \$8 $\times$ 2     **D.** \$11.50 + \$8 $\times$ 2 $\times$ 4

    **7.** _____

**8.** What is the value of $a + b - c$ if $a = 20$, $b = 10$, and $c = 5\frac{1}{2}$?

    **F.** $25$     **G.** $24\frac{1}{2}$     **H.** $35\frac{1}{2}$     **I.** $40$

    **8.** _____

**9.** Ambu has saved \$56 of her allowance money to buy books. If she buys 6 books at $d$ dollars per book, she will have $56 - 6d$ of her allowance left. How much does she have left if the books cost \$4.75 each?

    **A.** \$27.50     **B.** \$28.50     **C.** \$50     **D.** \$51.25

    **9.** _____

**Which is the correct algebraic expression for each phrase?**

**10.** 38 dollars less than Charlie earned

    **F.** $c + 38$     **G.** $38 - c$     **H.** $c - 38$     **I.** $38 + c$

    **10.** _____

**11.** 3.1 times as many meters

    **A.** $m + 3.1$     **B.** $3.1 + m$     **C.** $3.1 \div m$     **D.** $3.1m$

    **11.** _____

**12.** one fourth the number of rocks Joyce found

    **F.** $j \times 4$     **G.** $j \div 4$     **H.** $4 \div j$     **I.** $j - 4$

    **12.** _____

# Test, Form 2B    *(continued)*

13. Hamza made 6 calls in one day. Each call cost the same amount of money. The next day he made a call that cost $4. Which expression represents the total cost of the calls Hamza made during the two days?

    **A.** $6 + 4c$    **B.** $6c + 4$    **C.** $4c + 6c$    **D.** $6 + 4 + c$

    13. _____

14. Which property is illustrated by the statement $4 + (9 + 12) = (4 + 9) + 12$?

    **F.** Associative  **G.** Identity    **H.** Distributive  **I.** Commutative

    14. _____

15. Which property is illustrated by the statement $9 \times 1 = 9$?

    **A.** Associative  **B.** Identity    **C.** Distributive  **D.** Commutative

    15. _____

16. Which shows how to find $9 \times 305$ mentally by using the Distributive Property?

    **F.** $9(300) + (5)$          **H.** $5(300 + 9)$

    **G.** $9(300) + 5(5)$         **I.** $9(300) + 9(5)$

    16. _____

17. Five friends each spent $9 on bowling games and $3.50 on shoe rentals. Which expression cannot be used to find the total amount the friends spent?

    **A.** $5(\$9) + 5(\$3.50)$          **C.** $5(\$9 + \$3.50)$

    **B.** $5(\$12.50)$                  **D.** $5(\$9)$

    17. _____

**Use the Distributive Property to rewrite each expression.**

18. $4(y + 8)$

    18. _____

19. $3(2 + h)$

    19. _____

20. Find the value of $14 \times 2 + 8 \div 4$.

    20. _____

21. Evaluate $hj + h$ if $h = 6$ and $j = 8$.

    21. _____

**Determine whether the two expressions are equivalent. If so, tell what property is applied. If not, explain why.**

22. $7 + (2 + 3)$ and $(7 + 2) + 3$

    22. _____

23. $18 \times 5$ and $5 \times 18$

    23. _____

**Simplify each expression.**

24. $4x + 8x + 6x$

    24. _____

25. $9(5x)$

    25. _____

26. $3(7x + 2y)$

    26. _____

27. Six friends went to a hockey game. The price of admission per person was $x. Four of the friends bought a soda for $2. Write and simplify an expression that represents the total cost.

    27. _____

# Test, Form 3A

**Write the correct answer in the blank at the right of each question.**

1. Write $\left(\frac{1}{6}\right)^3$ as a product of the same factor. Then find the value.

1. _____

2. Write $9 \times 9 \times 9 \times 9 \times 9 \times 9$ using an exponent.

2. _____

**Find the value of each expression.**

3. $5 + 4 \times 5 - 7$

3. _____

4. $8 \times (3^3 + 2) - 11$

4. _____

5. $30 \times 4 - 2^2 \times 5$

5. _____

6. $55 \div 5 \times 2^3 \div 4$

6. _____

7. Tanya purchased 4 hammers for $11.79 each and 7 screwdrivers for $6.65 each. Write an expression for the total cost of the tools. Then find the total cost.

7. _____

**Evaluate each expression if $a = 3$, $b = \frac{2}{3}$, and $c = 6$.**

8. $2a + 5$

8. _____

9. $2c + 3a$

9. _____

10. $c^2 + 3a \times b$

10. _____

11. Koby took $43 to the movie theater. If he paid for 3 tickets that each cost $d$ dollars, he will have $43 - 3d$ left. How much does he have left if the tickets cost $9.75 each?

11. _____

**Write each phrase as an algebraic expression.**

12. 16 seconds faster than Najib's time

12. _____

13. one piece more than twice the number of pieces

13. _____

14. one third of Danielle's height

14. _____

15. six meters less than four times the width

15. _____

16. Devante made and sold 8 pitchers of lemonade at his lemonade stand. He used the same number of lemons in each pitcher. He had 3 lemons left over. Write an expression to represent the total number of lemons Devante had in all.

16. _____

17. Helen divided her colored pencils evenly among herself and three friends. Write an expression to represent the number of pencils each person received.

17. _____

# Test, Form 3A    *(continued)*

**Determine whether the two expressions are equivalent. If so, tell what property is applied. If not, explain why.**

**18.** $(8 \times 2) \times 5 = 8 \times (2 \times 5)$

18. _____

**19.** $13 + 0 = 13$

19. _____

**20.** $12 - (5 - 3) = (12 - 5) - 3$

20. _____

**21.** Niko worked out for 45 minutes, 22 minutes, and 25 minutes last week. Use the Commutative Property to find the total number of minutes he worked out.

21. _____

**Find each product mentally. Show the steps you used.**

**22.** $6 \times 52$

22. _____

**23.** $4 \times 7.1$

23. _____

**Use the Distributive Property to rewrite each algebraic expression.**

**24.** $5(x + 9)$

24. _____

**25.** $11(12 + r)$

25. _____

**26.** $9(v + 2)$

26. _____

**27.** $7(b + 3.5)$

27. _____

**28.** Mrs. Sosa bought a T-shirt and a pair of socks for each of her 14 grandchildren. The table lists the cost of each item. Use the Distributive Property to find the total amount of money she spent on her grandchildren.

| Item | Cost($) |
|--------|---------|
| Pants | 19.00 |
| Socks | 4.95 |
| T-shirt | 12.00 |

28. _____

**Simplify each expression.**

**29.** $7x + 3x + 4x$

29. _____

**30.** $5(2x + 4y)$

30. _____

**31.** $6x + 2y + 9x$

31. _____

**Factor each expression.**

**32.** $9x + 36y$

32. _____

**33.** $12x + 18y$

33. _____

# Test, Form 3B

**Write the correct answer in the blank at the right of each question.**

1. Write $\left(\dfrac{1}{9}\right)^3$ as a product of the same factor. Then find the value.

1. _____

2. Write $4 \times 4 \times 4 \times 4 \times 4 \times 4 \times 4 \times 4$ using an exponent.

2. _____

**Find the value of each expression.**

3. $8 + 17 - 5 \times 3$

3. _____

4. $7 + (2^4 + 7) \times 10$

4. _____

5. $3^2 \times 5 - 3 \times 2$

5. _____

6. $9 + 4^2 \div 8 \times 3$

6. _____

7. Elisa purchased 5 balls of yarn for $5.35 each and 9 jars of glue for $3.45 each. Write an expression for the total cost of the supplies. Then find the total cost.

7. _____

**Evaluate each expression if $x = 2$, $y = \dfrac{3}{5}$, and $z = 5$?**

8. $3z + 4$

8. _____

9. $5x + 2z$

9. _____

10. $x^2 + 5y \div z$

10. _____

11. Kenichi took $72 to a concert. If he paid for 5 tickets that each cost $d$ dollars, he will have $72 - 5d$ left. How much does he have left if the tickets cost $12.65 each?

11. _____

**Write each phrase as an algebraic expression.**

12. one fourth the amount of salt

12. _____

13. 30 seconds slower than Godfrey's time

13. _____

14. three pretzels more than twice the number of pretzels

14. _____

15. eight centimeters less than three times the height

15. _____

16. Alexia used three reams of paper in her first semester of school. Each ream has the same number of sheets. She used another 40 sheets after that. Write an expression to represent the total number of sheets Alexia used.

16. _____

17. Lester divided his grapes evenly among himself and four friends. Write an expression to represent the number of grapes each person received.

17. _____

# Test, Form 3B (continued)

**Determine whether the two expressions are equivalent. If so, tell what property is applied. If not, explain why.**

**18.** $18 \times 1 = 18$

18. _____

**19.** $8 + 1.2 = 1.2 + 8$

19. _____

**20.** $10 \div (8 \div 4) = (10 \div 8) \div 4$

20. _____

**21.** Sophia played the piano for 18 minutes and 26 minutes last week. Troy played the piano for 14 minutes last week. Use the Associative Property to find the total number of minutes they played.

21. _____

**Find each product mentally. Show the steps you used.**

**22.** $9 \times 34$

22. _____

**23.** $4 \times 8.2$

23. _____

**Use the Distributive Property to rewrite each algebraic expression.**

**24.** $3(y + 10)$

24. _____

**25.** $13(14 + t)$

25. _____

**26.** $8(a + 4)$

26. _____

**27.** $3(w + 1.6)$

27. _____

**28.** Mr. Lang bought a hat and a pair of gloves for each of his 12 grandchildren. The table lists the cost of each item. Use the Distributive Property to find the total amount of money he spent on his grandchildren.

| Item | Cost($) |
|------|---------|
| Coat | 30.00 |
| Gloves | 6.49 |
| Hat | 8.00 |

28. _____

**Simplify each expression.**

**29.** $8x + 6x + 2x$

29. _____

**30.** $4(3x + 7y)$

30. _____

**31.** $12y + 5x + 8y$

31. _____

**Factor each expression.**

**32.** $42x + 12y$

32. _____

**33.** $15x + 30y$

33. _____

# Are You Ready?

## Review

### Example

Find $\frac{5}{6} - \frac{3}{4}$.

The LCD of $\frac{5}{6}$ and $\frac{3}{4}$ is 12.

| Write the problem. | Rename using the LCD, 12. | Subtract the fractions. |
|---|---|---|
| $\frac{5}{6} \longrightarrow$ | $\frac{5 \times 2}{6 \times 2} = \frac{10}{12}$ | $\frac{10}{12}$ |
| $-\frac{3}{4} \longrightarrow$ | $\frac{3 \times 3}{4 \times 3} = \frac{9}{12}$ | $-\frac{9}{12}$ |
| | | $\frac{1}{12}$ |

### Exercises

**Solve. Write in simplest form.**

1. $\frac{5}{6} - \frac{2}{3}$

1. _____

2. $\frac{7}{8} - \frac{1}{4}$

2. _____

3. $\frac{7}{10} - \frac{2}{5}$

3. _____

4. **RUNNING** On Friday, Melvin ran $\frac{3}{8}$ mile, and on Saturday he ran $\frac{3}{10}$ mile. How much farther did Melvin run on Friday than on Saturday?

4. _____

5. $\frac{7}{9} - \frac{1}{3}$

5. _____

6. $\frac{2}{3} - \frac{4}{7}$

6. _____

7. $\frac{7}{10} - \frac{1}{2}$

7. _____

8. **STEW** Merlin made $\frac{7}{8}$ gallon of stew. Then he ate $\frac{3}{4}$ gallon of stew. How much stew was left over?

8. _____

# Are You Ready?

## Practice

**Find each difference.**

1. 5.78 − 2.42

2. 1.05 − 0.75

3. 4.5 − 2.95

4. 3.20 − 1.83

5. **PIES** At a local bakery cherry pies cost $8.75 and blueberry pies cost $10.50. How much more do blueberry pies cost than cherry pies?

**Subtract. Write in simplest form.**

6. $\frac{5}{6} - \frac{1}{3}$

7. $\frac{3}{4} - \frac{4}{9}$

8. $\frac{3}{4} - \frac{3}{10}$

9. $\frac{7}{10} - \frac{2}{5}$

10. **MAZE** It took Rattelda, the rat, $\frac{5}{6}$ of a minute to run through a maze. It took Mercury, the mouse, $\frac{2}{3}$ of a minute to run through the same maze. How much faster was the rat than the mouse?

11. **JUICE** On Tuesday, an orange juice bottle contained $\frac{7}{8}$ gallon of juice. On Friday, it contained $\frac{1}{6}$ gallon of juice. How much of the juice was consumed from Tuesday through Friday?

1. _____

2. _____

3. _____

4. _____

5. _____

6. _____

7. _____

8. _____

9. _____

10. _____

11. _____

# Are You Ready?

## Apply

1. **FARMING** A farmer plows $\frac{2}{3}$ of his field in the morning and then $\frac{1}{6}$ of his field in the afternoon. How much more of the field did he mow in the morning?

2. **TOURING** On a guided tour of a cave, a guide walks for a total of $\frac{3}{4}$ of an hour and stops and talks for a total of $\frac{2}{5}$ of an hour. How much longer was the walking portion than the talking portion of the tour?

3. **CHEF HAT** Brynn made a chef's hat for her father. The top of the hat was made with $\frac{6}{7}$ yard of linen and the brim of the hat was made with $\frac{2}{5}$ yard of linen. How much more linen was used for the top of the hat than the brim?

4. **BIRD FEEDER** A bird feeder held $\frac{7}{8}$ cup of birdseed until a flock of Northern Mockingbirds ate $\frac{1}{2}$ cup of seed. How much seed was left in the birdfeeder after that?

5. **WALK-A-THON** After one hour of walking, Sayaka had $\frac{4}{5}$ of the walk completed while Takara had $\frac{7}{10}$ of the walk completed. How much more of the walk had Sayaka completed than Takara?

6. **COOK-OFF** For a dip recipe, Atrina used $\frac{5}{6}$ of a pound of cream cheese. Kendra used $\frac{2}{3}$ of a pound of cream cheese. How much less cream cheese did Kendra use?

# Diagnostic Test

**Find each difference.**

1. $4.98 - 1.45$

2. $1.25 - 0.99$

3. $2.47 - 1.35$

4. $4.67 - 1.39$

5. **PIZZA** At the concession stand, a slice of cheese pizza costs $1.75 and a slice of pepperoni costs $2.20. How much more does a slice of pepperoni cost than a slice of cheese?

**Subtract. Write in simplest form.**

6. $\frac{3}{8} - \frac{1}{6}$

7. $\frac{3}{4} - \frac{3}{8}$

8. $\frac{5}{6} - \frac{1}{2}$

9. $\frac{11}{15} - \frac{2}{3}$

10. **RECREATION** Suppose $\frac{1}{3}$ of a group of students goes rollerblading while $\frac{2}{5}$ of the group plays beach volleyball. How much more of the group chooses to play beach volleyball than go rollerblading?

11. **TRAILS** The blue trail is $\frac{9}{10}$ mile long and the yellow trail is $\frac{7}{8}$ mile long. How much longer is the blue trail than the yellow trail?

1. _____

2. _____

3. _____

4. _____

5. _____

6. _____

7. _____

8. _____

9. _____

10. _____

11. _____

# Pretest

**Identify the solution of each equation from the list given.**

1. $b + 8 = 14$; 4, 5, 6

2. $25 - t = 15$; 9, 10, 11

**Solve each equation mentally.**

3. $9p = 81$

4. $\dfrac{y}{7} = 2$

**Solve each equation. Check your solution.**

5. $y + 6 = 10$

6. $f - 4 = 9$

7. $8d = 48$

8. $10 = \dfrac{r}{5}$

9. $15 = \dfrac{x}{3}$

10. **HAIR** Anna's hair is 24 inches long. This is 12 inches longer than her sister's. Write and solve an addition equation to find the length of her sister's hair.

11. **SPELLING** Nico had 9 words correct on his weekly spelling test. This is 1 less than what he had correct the week before. Write and solve a subtraction equation to find the number of spelling words Nico had correct the week before.

12. **CHARITY** This year, James raised $95 for an animal shelter. This is 5 times the amount he saved. The equation $5x = 95$ can be used to find last year's fundraising amount. Find last year's fundraising amount.

1. _____

2. _____

3. _____

4. _____

5. _____

6. _____

7. _____

8. _____

9. _____

10. _____

11. _____

12. _____

# Chapter Quiz

**Identify the solution of each equation from the list given.**

1. $b + 12 = 16$; 2, 3, 4

1. _____

2. $23 = 30 - g$; 6, 7, 8

2. _____

3. $6w = 48$; 6, 7, 8

3. _____

**Solve each equation mentally.**

4. $8p = 72$

4. _____

5. $\frac{y}{7} = 4$

5. _____

6. **SHOPPING** Ramos went to a pizza shop and spent half of his money on lunch. If Ramos had $4.50 left after lunch, how much money did he have originally?

6. _____

**Solve each equation. Check your solution.**

7. $y + 7 = 18$

7. _____

8. $h - 6.5 = 15$

8. _____

9. $24 = m - 15$

9. _____

10. **CARDS** In a game, Halley has 35 cards. This is 12 more cards than Felix. Write and solve an addition equation to find the number of cards Felix has.

10. _____

11. Billy is thinking of three numbers from 1 through 9 with a sum of 17. Each number is used only once. What are the numbers?

11. _____

# Vocabulary Test

| | | |
|---|---|---|
| Addition Property of Equality | equation | solution |
| Division Property of Equality | inverse operations | solve |
| equals sign | Multiplication Property of Equality | Subtraction Property of Equality |

## Fill in the blank to complete each sentence.

1. A(n)_____ is used in a mathematical sentence to show that two expressions are equal.

   1. _____

2. The _____ of $2x = 10$ is 5.

   2. _____

3. The _____ Property of Equality states that if you divide both sides of an equation by the same nonzero number, the two sides remain equal.

   3. _____

4. Multiplication and division are _____.

   4. _____

5. The _____ Property of Equality states that you can add the same number to both sides of an equation, and both sides will remain equal.

   5. _____

6. A(n) _____ must contain an equals sign.

   6. _____

7. Addition and subtraction are _____.

   7. _____

## In your own words, define the terms.

8. Subtraction Property of Equality

   8. _____

9. solve

   9. _____

# Standardized Test Practice

**Read each question. Then fill in the correct answer on the answer sheet provided by your teacher or on a sheet of paper.**

1. The receipt shows the quantity and price of some clothing. Which equation can be used to find the price of each pair of jeans?

| Casual Fashions | | |
|---|---|---|
| Quantity | Item | Price |
| 2 | pairs of jeans | $64.52 |
| 3 | sweaters | ▪ |
| | | |
| Total | | $136.67 |

   **A.** $64.52 + x = 136.67$
   **B.** $3x = 136.67$
   **C.** $64.52 = 2x$
   **D.** $2 + x = 64.52$

2. A sub shop is keeping track of the number of pounds of turkey sold each day.

| Day | Number of Pounds of Turkey Sold |
|---|---|
| Monday | 40.6 |
| Tuesday | 25.4 |
| Wednesday | 34.7 |
| Thursday | 45.2 |
| Friday | 65.3 |
| Saturday | 70.4 |
| Sunday | 50.8 |

   The amount of turkey sold on Saturday is about how many times the amount of turkey sold on Wednesday?

   **F.** about 1.5 times     **H.** about 3 times
   **G.** about 2 times       **I.** about 4 times

3. ▦✎ **GRIDDED RESPONSE** Bob's Boot Shop sold 60% of its stock of winter boots before the first snow of the year. What fraction of the stock of winter boots has not yet been sold?

4. Two thirds of a blueberry pie is left in the refrigerator. If the leftover pie is cut in 6 equal-size slices, which fraction of the original pie is each slice?

   **A.** $\frac{1}{9}$          **C.** $\frac{1}{4}$

   **B.** $\frac{1}{6}$          **D.** $\frac{1}{3}$

5. ▦✎ **GRIDDED RESPONSE** Mikayla bought a value pack of crackers for $6.72. The value pack had 24 individually wrapped cracker packages. Solve $24x = 6.72$ to find the cost per package in dollars.

6. At a pet store, there are 30 aquariums. Of the 30, 20 have fresh water with 15 fish each. The other 10 aquariums have salt water with 6 fish each. Which equation could be used to find $f$, the total number of fish at the pet store?

   **F.** $f = 30 + 20 + 15 + 10 + 6$
   **G.** $f = 30(15 + 6)$
   **H.** $f = 20(15 + 6)$
   **I.** $f = (20 \times 15) + (10 \times 6)$

7. ▦✎ **GRIDDED RESPONSE** The ratio of boys to girls at a rock concert was about 7 to 6. If there were 1,092 boys at the rock concert, how many girls were there?

8. What value of $x$ makes this equation true?

$$4x = 44$$

   **A.** 14          **C.** 22

   **B.** 13          **D.** 11

**9.** [THINK SOLVE EXPLAIN] **SHORT RESPONSE** A movie theater has 26 rows of seats with an equal number of seats in each row. The theater can seat a total of 390 people. Solve the equation $26x = 390$ to find the number of seats $x$ that are in each row.

**10.** Which of the following diagrams represents 80%?

**F.**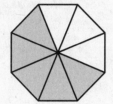

**H.**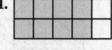

**G.**

**I.**

**11.** The rates for renting a jet ski are shown.

| East Lake Rentals | |
|---|---|
| Weekday cost per hour | $95 |
| Weekend cost per hour | $110 |

Which expression could be used to find the cost of renting a jet ski for 3 hours on Friday and 2 hours on Saturday?

**A.** $5(95 + 110)$

**B.** $3(95) + 2(110)$

**C.** $3(95) \times 2(110)$

**D.** $(3 + 95) \times (2 + 110)$

**12.** [THINK SOLVE EXPLAIN] **SHORT RESPONSE** During a vacation, Alice went fly-fishing. She rented a driftboat.

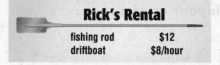

**Rick's Rental**

| | |
|---|---|
| fishing rod | $12 |
| driftboat | $8/hour |

If she spent a total of $96, write and solve an equation to find how many hours she rented the boat.

**13.** [THINK SOLVE EXPLAIN] **EXTENDED RESPONSE** Mr. Martin and his son are planning a fishing trip to Alaska. The rates of two companies are shown.

**Wild Fishing, Inc.**
Per Person Rates
Round-Trip Flight $200 / Daily Rate $25

**Fisherman's Service**
Per Person Rates
Round-Trip Flight $150 / Daily Rate $30

*Part A* Which company offers a better deal for a three-day trip for Mr. Martin and his son? What will be the cost?

*Part B* Which company would charge $700 for 6 days for both of them?

*Part C* Write an expression to represent the cost of the trip for Mr. Martin and his son for each company.

NAME _____ DATE _____ PERIOD _____

# Student Recording Sheet

SCORE _____

Use this recording sheet with the Standardized Test Practice pages.

**Fill in the correct answer. For gridded-response questions, write your answers in the boxes on the answer grid and fill in the bubbles to match your answers.**

1. Ⓐ Ⓑ Ⓒ Ⓓ

2. Ⓕ Ⓖ Ⓗ Ⓘ

3.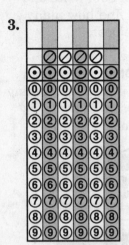

4. Ⓐ Ⓑ Ⓒ Ⓓ

5.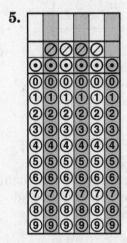

6. Ⓕ Ⓖ Ⓗ Ⓘ

7.

8. Ⓐ Ⓑ Ⓒ Ⓓ

9. _____

10. Ⓕ Ⓖ Ⓗ Ⓘ

11. Ⓐ Ⓑ Ⓒ Ⓓ

12. _____

## Extended Response
Record your answers for Exercise 13 on the back of this paper.

# Extended-Response Test

Demonstrate your knowledge by giving a clear, concise solution to each problem. Be sure to include all relevant drawings and justify your answers. You may show your solution in more than one way or investigate beyond the requirements of the problem. If necessary, record your answer on another piece of paper.

**1.** Write and solve an equation for the following situation.

**CLASSES** Five students each paid $m$ dollars to enroll in a community college program. They paid a total of $975. How much does the program cost per student?

**Use the table to answer Exercises 2–4.**

**BEARS** A local shop called Stuff-4-Fun allows children to come in and purchase bears that they can stuff and then dress in particular outfits and accessories. It costs $x$ dollars to buy a bear at the Stuff-4-Fun shop, plus the cost on any additional items.

| Additional Items | Cost |
|------------------|--------|
| Cheerleader outfit | $12.50 |
| Football uniform | $10.50 |
| Voice box | $15.00 |
| Soccer uniform | $9.50 |

**2.** Carter bought a bear and paid for a football uniform. The total cost was $38.50. Write and solve an equation to find the cost of buying a bear.

**3.** Each girl on a cheerleading squad bought a bear. The leader paid for cheerleading outfits for each of them. She spent $175. Write and solve an equation to find how many cheerleading outfits the leader bought.

**4.** Mrs. Hernandez bought several soccer uniforms. She spent $57 in all. Write and solve an equation to find the number of soccer uniforms she bought.

# Extended-Response Rubric

| Score | Description |
|:---:|---|
| 4 | A score of four is a response in which the student demonstrates a thorough understanding of the mathematics concepts and/or procedures embodied in the task. The student has responded correctly to the task, used mathematically sound procedures, and provided clear and complete explanations and interpretations.<br><br>The response may contain minor flaws that do not detract from the demonstration of a thorough understanding. |
| 3 | A score of three is a response in which the student demonstrates an understanding of the mathematics concepts and/or procedures embodied in the task. The student's response to the task is essentially correct with the mathematical procedures used and the explanations and interpretations provided demonstrating an essential but less than thorough understanding.<br><br>The response may contain minor flaws that reflect inattentive execution of mathematical procedures or indications of some misunderstanding of the underlying mathematics concepts and/or procedures. |
| 2 | A score of two indicates that the student has demonstrated only a partial understanding of the mathematics concepts and/or procedures embodied in the task. Although the student may have used the correct approach to obtaining a solution or may have provided a correct solution, the student's work lacks an essential understanding of the underlying mathematical concepts.<br><br>The response contains errors related to misunderstanding important aspects of the task, misuse of mathematical procedures, or faulty interpretations of results. |
| 1 | A score of one indicates that the student has demonstrated a very limited understanding of the mathematics concepts and/or procedures embodied in the task. The student's response is incomplete and exhibits many flaws. Although the student's response has addressed some of the conditions of the task, the student reached an inadequate conclusion and/or provided reasoning that was faulty or incomplete.<br><br>The response exhibits many flaws or may be incomplete. |
| 0 | A score of zero indicates that the student has provided no response at all, or a completely incorrect or uninterpretable response, or demonstrated insufficient understanding of the mathematics concepts and/or procedures embodied in the task. For example, a student may provide some work that is mathematically correct, but the work does not demonstrate even a rudimentary understanding of the primary focus of the task. |

# Test, Form 1A

**Write the letter for the correct answer in the blank at the right of each question.**

**What is the solution of each equation?**

1. $9 + k = 18$
   - **A.** 18
   - **B.** 9
   - **C.** 8
   - **D.** 7

   1. _____

2. $r - 11 = 5$
   - **F.** 11
   - **G.** 14
   - **H.** 15
   - **I.** 16

   2. _____

3. $3w = 30$
   - **A.** 6
   - **B.** 9
   - **C.** 10
   - **D.** 12

   3. _____

4. $\frac{d}{12} = 4$
   - **F.** 2
   - **G.** 3
   - **H.** 8
   - **I.** 48

   4. _____

5. $y + 5 = 10$
   - **A.** 15
   - **B.** 10
   - **C.** 6
   - **D.** 5

   5. _____

6. $13 = t + 7$
   - **F.** 5
   - **G.** 6
   - **H.** 7
   - **I.** 20

   6. _____

7. $a - 5 = 9$
   - **A.** 14
   - **B.** 9
   - **C.** 5
   - **D.** 4

   7. _____

8. $10 = t - 8$
   - **F.** 2
   - **G.** 4
   - **H.** 18
   - **I.** 20

   8. _____

9. Janeen's mother is 47. She is 26 years older than Janeen. Which equation can be used to find Janeen's age $j$?
   - **A.** $j + 26 = 47$
   - **B.** $26j = 47$
   - **C.** $j - 26 = 47$
   - **D.** $47j = 26$

   9. _____

10. Quentin bought 6 new tennis balls. If he now has a total of 18 tennis balls, how many did he start with?
    - **F.** 9
    - **G.** 12
    - **H.** 21
    - **I.** 24

    10. _____

# Test, Form 1A   *(continued)*

**What is the solution of each equation?**

**11.** $6d = 24$

    **A.** 4         **B.** 5         **C.** 18         **D.** 144     **11.** _____

**12.** $42 = 7f$

    **F.** 294       **G.** 49         **H.** 35         **I.** 6           **12.** _____

**13.** $4 = \dfrac{x}{8}$

    **A.** 2         **B.** 4         **C.** 32         **D.** 48       **13.** _____

**14.** $\dfrac{t}{5} = 15$

    **F.** 100       **G.** 75         **H.** 10         **I.** 3           **14.** _____

**15.** Antwaun waters his lawn 3 times a week. If he watered his lawn 24 times in all, which equation could be used to find how many weeks he has been watering his lawn?

    **A.** $x + 3 = 24$         **C.** $3x = 24$

    **B.** $24 - x = 3$         **D.** $\dfrac{x}{3} = 24$         **15.** _____

**16.** At a nursery, plants are half off their original price. The sale price of a potted plant is \$5.50. Which equation could be used to find the original cost of the plant?

    **F.** $\dfrac{c}{2} = \$5.50$         **H.** $\$5.50c = 2$

    **G.** $2c = \$5.50$         **I.** $2 + c = \$5.50$       **16.** _____

**What is the solution of each equation?**

**17.** $\dfrac{x}{4} = 12$

    **A.** 3         **B.** 8         **C.** 48         **D.** 52       **17.** _____

**18.** $4f = 44$

    **F.** 6         **G.** 8         **H.** 11         **I.** 96       **18.** _____

**19.** $9b = 54$

    **A.** 558       **B.** 486      **C.** 6.8       **D.** 6         **19.** _____

**20.** Kelly earned \$20 for babysitting. She has also earned money by doing chores. Altogether, she has earned \$100. How much did she earn doing chores?

    **F.** \$5        **G.** \$20      **H.** \$80      **I.** \$120     **20.** _____

# Test, Form 1B

SCORE _____

**Write the letter for the correct answer in the blank at the right of each question.**

**What is the solution of each equation?**

1. $8 + g = 18$
   
   **A.** 26      **B.** 11      **C.** 10      **D.** 8

   1. _____

2. $p - 10 = 5$
   
   **F.** 11      **G.** 14      **H.** 15      **I.** 16

   2. _____

3. $6w = 60$
   
   **A.** 6      **B.** 10      **C.** 12      **D.** 30

   3. _____

4. $\frac{d}{14} = 7$
   
   **F.** 2      **G.** 7      **H.** 21      **I.** 98

   4. _____

5. $y + 7 = 11$
   
   **A.** 18      **B.** 17      **C.** 5      **D.** 4

   5. _____

6. $13 = t + 2$
   
   **F.** 15      **G.** 13      **H.** 11      **I.** 9

   6. _____

7. $a - 5 = 12$
   
   **A.** 5      **B.** 7      **C.** 17      **D.** 60

   7. _____

8. $13 = t - 4$
   
   **F.** 52      **G.** 17      **H.** 9      **I.** 8

   8. _____

9. A pet store has 45 mice. The number of mice is 22 more than the number of cats. Which equation can be solved to find the number of cats $c$?
   
   **A.** $22c = 45$          **C.** $c + 22 = 45$
   
   **B.** $45c = 22$          **D.** $c - 22 = 45$

   9. _____

10. Tulia bought 9 new golf balls. If she now has a total of 22 golf balls, how many did she start with?
    
    **F.** 31      **G.** 26      **H.** 13      **I.** 9

    10. _____

# Test, Form 1B   *(continued)*

**What is the solution of each equation?**

**11.** $7d = 35$

    **A.** 5         **B.** 6         **C.** 42         **D.** 245         **11.** _____

**12.** $42 = 6f$

    **F.** 7         **G.** 8         **H.** 48         **I.** 252         **12.** _____

**13.** $9 = \frac{x}{7}$

    **A.** 16         **B.** 54         **C.** 56         **D.** 63         **13.** _____

**14.** $\frac{e}{4} = 20$

    **F.** 5         **G.** 8         **H.** 80         **I.** 100         **14.** _____

**15.** Four friends bought tickets to the school play. If it cost them a total of \$28, which equation could be used to find how much each friend paid?

    **A.** $\frac{x}{4} = 28$         **C.** $x + 4 = 28$

    **B.** $28 - x = 4$         **D.** $4x = 28$         **15.** _____

**16.** The height reached by Amal's rocket was one fourth the height reached by Min's rocket. Amal's rocket reached a height of 15 meters. Which equation could be used to find the height reached by Min's rocket?

    **F.** $4m = 15$         **H.** $\frac{m}{4} = 15$

    **G.** $15m = 4$         **I.** $4 + m = 15$         **16.** _____

**What is the solution of each equation?**

**17.** $\frac{x}{5} = 15$

    **A.** 3         **B.** 10         **C.** 20         **D.** 75         **17.** _____

**18.** $4b = 32$

    **F.** 128         **G.** 11         **H.** 10.5         **I.** 8         **18.** _____

**19.** $8c = 40$

    **A.** 5         **B.** 6         **C.** 32         **D.** 40         **19.** _____

**20.** Chad earned \$30 for raking leaves. He has also earned money by doing chores. Altogether, he has earned \$120. How much did he earn doing chores?

    **F.** \$15         **G.** \$20         **H.** \$90         **I.** \$150         **20.** _____

# Test, Form 2A

**Write the letter for the correct answer in the blank at the right of each question.**

**What is the solution of each equation?**

**1.** $18 + g = 38$
  **A.** 56     **B.** 30     **C.** 20     **D.** 12

1. _____

**2.** $p - 16 = 24$
  **F.** 40     **G.** 34     **H.** 30     **I.** 8

2. _____

**3.** $\frac{d}{75} = 25$
  **A.** $\frac{1}{3}$     **B.** 3     **C.** 500     **D.** 1,875

3. _____

**4.** $13 = t + 2.1$
  **F.** 15.9     **G.** 15.1     **H.** 11.9     **I.** 10.9

4. _____

**5.** $y + 174 = 200$
  **A.** 374     **B.** 126     **C.** 36     **D.** 26

5. _____

**6.** $y + \frac{1}{2} = \frac{3}{4}$
  **F.** $1\frac{1}{2}$     **G.** $1\frac{1}{4}$     **H.** $\frac{1}{2}$     **I.** $\frac{1}{4}$

6. _____

**7.** $13.2 = t - 4.4$
  **A.** 176     **B.** 88     **C.** 17.6     **D.** 8.8

7. _____

**8.** Seema runs 5 miles more per week than her sister Padma. If Seema runs 17 miles per week, which equation could be used to find the number of miles Padma runs each week?
  **F.** $p + 17 = 5$        **H.** $p + 5 = 17$
  **G.** $p - 17 = 5$        **I.** $p - 5 = 17$

8. _____

**9.** Hugh bought 2 bottles of water. If the total cost of the water was $4.50, which equation could be used to find the cost of each bottle of water?
  **A.** $\$4.50x = 2$        **C.** $\frac{x}{\$4.50} = 2$
  **B.** $\$4.50 - x = 2$     **D.** $2x = \$4.50$

9. _____

# Test, Form 2A   *(continued)*

SCORE _____

10. One third of the tomatoes in a garden are ripe. If 6 tomatoes are ripe, which equation can be used to find the total number of tomatoes in the garden?

    **F.** $\frac{t}{3} = 6$     **G.** $t - 6 = 3$     **H.** $3t = 6$     **I.** $6t = 3$

    10. _____

**What is the solution of each equation?**

11. $18 = \frac{2}{5}f$

    **A.** $7\frac{1}{5}$     **B.** $7\frac{1}{2}$     **C.** 40     **D.** 45

    11. _____

12. $27 = \frac{z}{0.3}$

    **F.** 0.81     **G.** 8.1     **H.** 81     **I.** 90

    12. _____

13. $5x = 15$

    **A.** 108     **B.** 72     **C.** 3     **D.** 2

    13. _____

14. $2s = 34$

    **F.** 7     **G.** 17     **H.** 20     **I.** 60

    14. _____

15. Mercedes is 132 centimeters tall. This is twice the height of her younger brother. How tall is her younger brother?

    **A.** 27 cm     **B.** 66 cm     **C.** 71 cm     **D.** 140 cm

    15. _____

16. Solve $a - 152 = 90$.

    16. _____

17. Solve $4.8d = 2.4$.

    17. _____

18. Solve $\frac{b}{14} = 21$.

    18. _____

19. For every quarter spent, Reggie receives 50 food pebbles to feed the fish at a hatchery. Reggie has thrown 350 food pebbles into the fish pools. Write and solve an equation to find the number of quarters Reggie spent at the hatchery.

    19. _____

20. An exterminator charges $80 to visit a home and $35 to spray each room. If the total bill was $395, how many rooms were sprayed?

    20. _____

# Test, Form 2B

**Write the letter for the correct answer in the blank at the right of each question.**

**What is the solution of each equation?**

1. $24 + p = 54$
   A. 78    B. 30    C. 20    D. 14

   1. _____

2. $w - 16 = 28$
   F. 12    G. 18    H. 34    I. 44

   2. _____

3. $\frac{e}{120} = 30$
   A. 3,600    B. 150    C. 4    D. $\frac{1}{4}$

   3. _____

4. $18 = t + 2.3$
   F. 20.3    G. 16.3    H. 15.7    I. 12.7

   4. _____

5. $q + 166 = 300$
   A. 134    B. 144    C. 366    D. 466

   5. _____

6. $y + \frac{1}{2} = \frac{7}{10}$

   F. $\frac{1}{5}$    G. $\frac{3}{4}$    H. $1\frac{1}{5}$    I. $1\frac{1}{2}$

   6. _____

7. $14.2 = t - 5.4$
   A. 0.88    B. 1.96    C. 8.8    D. 19.6

   7. _____

8. Marco's model bridge is 9 centimeters taller than Sho's model bridge. If Sho's bridge is 31 centimeters tall, which equation could be used to find the height of Marco's bridge?
   F. $m + 9 = 31$       H. $m + 31 = 9$
   G. $m - 9 = 31$       I. $31 - 9 = m$

   8. _____

9. Anne bought 3 hats for $19.50. Which equation could be used to find the cost of each hat?

   A. $3x = \$19.50$       C. $\$19.50x = 3$
   B. $\$19.50 - x = 3$       D. $\frac{x}{\$19.50} = 3$

   9. _____

# Test, Form 2B   (continued)

**10.** One fifth of the plants in a vegetable garden are cucumber plants. If there are 8 cucumber plants, which equation can be used to find the total number of plants in the garden?

**F.** $5p = 8$     **G.** $p - 5 = 8$    **H.** $\frac{p}{5} = 8$     **I.** $8p = 5$

10. _____

**What is the solution of each equation?**

**11.** $16 = \frac{2}{3}f$

  **A.** 48        **B.** 32        **C.** 24        **D.** $10\frac{2}{3}$

11. _____

**12.** $24 = \frac{z}{0.6}$

  **F.** 4        **G.** 14.4        **H.** 24.6        **I.** 40

12. _____

**13.** $4x = 24$

  **A.** 6        **B.** 10        **C.** 54        **D.** 90

13. _____

**14.** $3s = 15$

  **F.** 27.9        **G.** 9.3        **H.** 6.9        **I.** 5

14. _____

**15.** Donte is 138 centimeters tall. This is three times the height of his dog. What is the height of his dog?

  **A.** 46 cm                 **C.** 135 cm

  **B.** 50 cm                 **D.** 150 cm

15. _____

**16.** Solve $r - 119 = 70$.

16. _____

**17.** Solve $4.8d = 1.2$.

17. _____

**18.** Solve $\frac{a}{12} = 32$.

18. _____

**19.** For every dime spent, Davida receives 15 food pebbles to feed the animals in the petting zoo. Davida has fed the animals a total of 135 food pebbles. Write and solve an equation to find the number of dimes Davida spent at the petting zoo.

19. _____

**20.** A cleaning company charges $125 to visit a home and $40 to clean each room. If the total bill was $365, how many rooms were cleaned?

20. _____

# Test, Form 3A

**Solve each equation mentally.**

1. $p + 75 = 100$

    1. _____

2. $p - 15 = 30$

    2. _____

3. $25v = 100$

    3. _____

4. Samantha brought a bucket of shells home from the beach. She gave 12 shells to her brother. If she now has 47 shells, how many shells did she bring home from the beach?

    4. _____

**Solve each equation. Check your solution.**

5. $y + 54 = 80$

    5. _____

6. $24.8 = t + 5.1$

    6. _____

7. $y + \frac{4}{9} = \frac{2}{3}$

    7. _____

8. $a - 45 = 91$

    8. _____

9. Enzo picked 56 strawberries at a farm. This is 25 more than the number of apples he picked. Write and solve an addition equation to find the number of apples Enzo picked.

    9. _____

10. Tina noticed that she had taken notes on 25 pages of her notebook. Ninety-five pages were still blank. Write and solve a subtraction equation to find the total number of pages in the notebook.

    10. _____

11. The difference between the high and low temperatures one day was 26°F. The low was 64°F. Write and solve a subtraction equation to find the high temperature.

    11. _____

# Test, Form 3A    (continued)

**Solve each equation. Check your solution.**

**12.** $12d = 216$

**12.** _____

**13.** $71.5 = 5.5u$

**13.** _____

**14.** $\dfrac{5}{6}x = \dfrac{5}{14}$

**14.** _____

**15.** $44 = \dfrac{t}{11}$

**15.** _____

**16.** $8.3 = \dfrac{b}{14}$

**16.** _____

**17.** $\dfrac{x}{3} = 15$

**17.** _____

**18.** $7.8x = 23.4$

**18.** _____

**19.** $53 = \dfrac{r}{4}$

**19.** _____

**20.** Mr. Miles spent $22.50 on rides at the carnival for his grandchildren. Each ride cost $1.25. Write and solve an equation to find how many rides he paid for in all.

**20.** _____

**21.** There were 196 students that participated in a clean up event at a nearby lake. This is one fourth of the entire student population. Write and solve an equation to find the number of students at this school.

**21.** _____

**22.** Mallory needs to go to the airport. It takes her 120 minutes to get there by car. This is 3 times the time it takes by train. Write and solve an equation to find how long it takes Mallory to get to the airport by train.

**22.** _____

**23.** For his book report, Bobby made color copies and black-and-white copies. The black-and-white copies cost $2.75 in all. The color copies cost $0.08 each. How many color copies did Bobby make if he spent $4.03 in all?

**23.** _____

# Test, Form 3B

**Solve each equation mentally.**

1. $k + 85 = 111$

1. _____

2. $f - 20 = 40$

2. _____

3. $12v = 36$

3. _____

4. Bethany bought a box of markers. She gave 6 markers to her friend. If Bethany now has 52 markers, how many markers were in the box she bought?

4. _____

**Solve each equation. Check your solution.**

5. $w + 62 = 90$

5. _____

6. $25.3 = s + 6.2$

6. _____

7. $y + \frac{3}{8} = \frac{3}{4}$

7. _____

8. $c - 39 = 82$

8. _____

9. Isiah earned 96 points on his math test. This is 11 points more than the number of points he earned on his history test. Write and solve an addition equation to find the number of points Isiah earned on his history test.

9. _____

10. After buying a jacket that cost $47, Marilyn had $6 left in her wallet. Write and solve a subtraction equation to find the amount of money Marilyn had in her wallet before she bought the jacket.

10. _____

11. The difference between the greatest and least weights of puppies in a shelter is 28 pounds. The lightest puppy weighs 4 pounds. Write and solve a subtraction equation to find the weight of the heaviest puppy.

11. _____

# Test, Form 3B   (continued)

**Solve each equation. Check your solution.**

**12.** $13g = 156$

12. _____

**13.** $99.2 = 6.2y$

13. _____

**14.** $\frac{3}{7}d = \frac{5}{14}$

14. _____

**15.** $52 = \frac{h}{12}$

15. _____

**16.** $7.6 = \frac{w}{12}$

16. _____

**17.** $\frac{x}{2} = 20$

17. _____

**18.** $8.3x = 41.5$

18. _____

**19.** $54 = \frac{r}{3}$

19. _____

**20.** Tomás earned \$38.25 for cleaning the garage. He was paid \$4.25 per hour. Write and solve an equation to find how many hours it took him to clean the garage.

20. _____

**21.** There were 214 students that voted for Jamison to be the student council president. This is one third of the number of students that voted. Write and solve an equation to find the number of students who voted.

21. _____

**22.** Emiko needs to go to the dentist. It takes her 70 minutes to get there if she rides her bike. This is twice the time it would take if she took the bus. Write and solve an equation to find how long it takes Emiko to get to the dentist by bus.

22. _____

**23.** Fred is making a bouquet of carnations and roses. The carnations cost \$5.25 in all. The roses cost \$1.68 each. How many roses did Fred use if the bouquet cost \$18.69 in all?

23. _____

# Are You Ready?

## Review

### Example 1

**Solve $63 + x = 140$.**

| | |
|---|---|
| $63 + x = 140$ | Write the equation. |
| $-63 \qquad -63$ | Subtract 63 to undo addition of 63. |
| $x = 77$ | Subtract. |

**Check** $63 + 77 = 140$ ✓

### Example 2

**Solve $16x = 64$.**

| | |
|---|---|
| $16x = 64$ | Write the equation. |
| $\dfrac{16x}{16} = \dfrac{64}{16}$ | Divide by 16 to undo multiplication by 16. |
| $x = 4$ | Divide. |

**Check** $16(4) = 64$ ✓

## Exercises

**Solve each equation.**

1. $n + 25 = 60$

2. $x - 10 = 18$

3. $p - 4 = 12$

4. $r + 7 = 15$

5. $13m = 39$

6. $19j = 38$

7. **COFFEE** A gourmet coffee shop bought 45 pounds of roasted coffee beans in two weeks. If they bought 20 pounds the second week, how many pounds did they buy the first week?

8. **LAUNDROMAT** A local laundromat uses 120 gallons of water in three days. How much water does it use each day if it uses the same amount every day?

1. _____

2. _____

3. _____

4. _____

5. _____

6. _____

7. _____

8. _____

# Are You Ready?

## Practice

**Replace ● with <, >, or = to make a true statement.**

1. 158,652 ● 185,652

2. 67,200 ● 672,000

3. 76,898 ● 72,898

4. 512,469 ● 512,649

5. **SAVINGS** Lasil's savings account had $408 last year. This year, it has $480. Compare $408 and $480 using <, >, or =.

6. **FACTORY** The automotive parts factory had 1,324 employees two years ago. They opened a new line and now have 1,423 employees. Compare 1,324 and 1,423 using <, >, or =.

**Solve each equation.**

7. $x + 36 = 78$     8. $8 + x = 24$

9. $x - 15 = 12$     10. $x - 9 = 10$

11. **CLOTHES** Josephina spent two days shopping for school clothes. She spent $115. If she spent $55 the first day, how much did she spend the second day?

**Solve each equation.**

12. $12p = 60$     13. $6n = 36$

14. $20r = 100$     15. $4n = 24$

16. **EXERCISE** Leo exercises for 60 minutes each day. He swims, runs, and lifts weights. How much time does he spend on each exercise if he spends the same amount of time on each?

1. _____

2. _____

3. _____

4. _____

5. _____

6. _____

7. _____

8. _____

9. _____

10. _____

11. _____

12. _____

13. _____

14. _____

15. _____

16. _____

# Are You Ready?

## Apply

---

**1. ARCHERY** The school wants to buy new bows for the archery team. One store charges $485 per bow. Another store charges $505. How much does the cheaper bow cost?

**2. BICYCLING** Adam rode his bike every day after school for one week. He rode 80 minutes in all. How much time did he spend riding daily if he rode the same amount of time each day?

---

**3. SCHOOL** The school collected donations for new equipment for three days. A total of $590 was donated. How much was donated on the second day?

| Day | Donations ($) |
|-----|---------------|
| 1 | 145 |
| 2 | ▓ |
| 3 | 280 |

**4. MONEY** An investor made $2,390 on Stock A. He made $2,250 on Stock B. Which stock made more money?

---

**5. SONGS** The school choir practices 4 songs for 60 minutes. How much time do they spend practicing each song if they spend the same amount of time on each?

**6. BASKETBALL** Basketball practice at the school gym lasts for 45 minutes. There are 15 students on the basketball team. How much time does the coach spend helping each member if he spends the same amount of time with each?

---

# Diagnostic Test

**Replace ● with <, >, or = to make a true statement.**

1. 468,523 ● 648,523

2. 895,000 ● 89,500

3. 36,542 ● 38,542

4. 972,314 ● 972,413

5. **TRUCKING** A trucker drove 214 miles one day and 241 miles the next day. Compare 214 and 241 using <, >, or =.

6. **COMPANY** A computer company sold 2,380 new computers last year. This year, they sold 2,208 new computers. Compare 2,380 and 2,208 using <, >, or =.

**Solve each equation.**

7. $x + 48 = 82$

8. $9 + x = 27$

9. $x - 11 = 14$

10. $x - 5 = 13$

11. **CARS** A car dealership sold 160 cars in two months. If they sold 86 cars in the second month, how many cars did they sell in the first month?

**Solve each equation.**

12. $18r = 54$

13. $7x = 49$

14. $17m = 34$

15. $9n = 72$

16. **PHONE** Chaz talked on the phone for 75 minutes total from Monday to Friday. If he talked for the same amount of time each day, how much time did he talk each day?

1. _____

2. _____

3. _____

4. _____

5. _____

6. _____

7. _____

8. _____

9. _____

10. _____

11. _____

12. _____

13. _____

14. _____

15. _____

16. _____

# Pretest

**1.** Complete the function table.

| Input (x) | x + 10 | Output (y) |
|-----------|--------|------------|
| 5 | 5 + 10 | |
| 8 | 8 + 10 | |
| 11 | 11 + 10 | |

**1.**

**2.** Find the value of the tenth term in the sequence.

| Position | 1 | 2 | 3 | 4 | 10 |
|----------|---|---|---|---|-----|
| Value of Term | 4 | 5 | 6 | 7 | |

**2.** _____

**3.** Graph $y = x + 2$.

**3.**

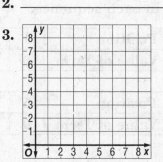

**4. SPAGHETTI** Plates at a spaghetti supper fundraiser cost $6 each. Write a function rule to find the cost of any number of plates $p$.

**4.** _____

## Is the given value a solution of the inequality?

**5.** $x + 5 > 7$, $x = 3$

**5.** _____

**6.** $x + 2 > 12$, $x = 5$

**6.** _____

**7.** $10 - x < 5$, $x = 7$

**7.** _____

**8.** Graph $n < 8$ on a number line.

**8.**

0   4   8   12   16

**9.** Solve $x + 3 \geq 9$.

**9.** _____

**10.** Solve $x - 1 \leq 11$.

**10.** _____

# Chapter Quiz

**1.** Complete the function table at the right.

**1.**

| Input, $x$ | $x - 4$ | Output, $y$ |
|---|---|---|
| 4 | | |
| 6 | | |
| 8 | | |

**Use words and symbols to describe the value of each term as a function of its position. Then find the twelfth term in each sequence.**

**2.**

| Position | 1 | 2 | 3 | 4 | $n$ |
|---|---|---|---|---|---|
| Value of Term | 3 | 6 | 9 | 12 | ▪ |

**2.** _____

**3.**

| Position | 5 | 6 | 7 | 8 | $n$ |
|---|---|---|---|---|---|
| Value of Term | 0 | 1 | 2 | 3 | ▪ |

**3.** _____

**Write an equation to represent the function.**

**4.** _____

**4.**

| Input, $x$ | 1 | 2 | 3 | 4 |
|---|---|---|---|---|
| Output, $y$ | 4 | 8 | 12 | 16 |

**5.**

| Input, $x$ | 0 | 1 | 2 | 3 |
|---|---|---|---|---|
| Output, $y$ | 0 | 12 | 24 | 36 |

**5.** _____

**Use the following information for Exercises 6–9.**

**SPORTS** In a football game, each team earns 6 points for each touchdown it scores.

**6.** Write an equation to find $y$, the total number of points for scoring $x$ touchdowns.

**6.** _____

**7.** Make a table to show the relationship between the number of touchdowns scored $x$ and the total points $y$ for 1, 2, and 3 touchdowns.

**7.**

| Touchdowns, $x$ | | | |
|---|---|---|---|
| Points, $y$ | | | |

**8.** Graph the ordered pairs $(x, y)$.

**8.**

**9.** How many points will a team earn if they score 7 touchdowns?

**9.** _____

# Vocabulary Test

| | | |
|---|---|---|
| arithmetic sequence | function table | linear function |
| dependent variable | geometric sequence | sequence |
| function | independent variable | term |
| function rule | inequality | |

## Choose the correct term to complete each sentence.

1. A (function, sequence) is a list of numbers in a specific order.

1. _____

2. A(n) (dependent, independent) variable is also known as the input value in a functions.

2. _____

3. A (linear, nonlinear) function has a graph that is a line.

3. _____

4. A (function rule, function table) organizes the input and output values.

4. _____

5. Each number in a sequence is called a (*y*-coordinate, term).

5. _____

6. A (function, sequence) is a relation that assigns one output value to one input value.

6. _____

7. The output value of a function is also known as the (dependent, independent) variable.

7. _____

## Define each term in your own words.

8. inequality

8. _____

9. arithmetic sequence

9. _____

# Standardized Test Practice

**Read each question. Then fill in the correct answer on the answer document provided by your teacher or on a sheet of paper.**

1. The table shows Molly and Myles's ages.

| Molly's Age, x (years) | Myles's Age, y (years) |
|---|---|
| 2 | 5 |
| 3 | 6 |
| 4 | 7 |
| 5 | 8 |

   Which expression **best** represents Myles's age in terms of Molly's age?

   **A.** $y + 3$

   **B.** $3x$

   **C.** $x + 3$

   **D.** $3y$

2. The cost of renting a speed boat is $25 plus an additional fee of $12 for each hour the boat is rented. Which equation can be used to find $c$, the cost for renting the boat for $h$ hours?

   **F.** $c = 12h + 25$

   **G.** $c = 25h + 12$

   **H.** $c = 12(h + 25)$

   **I.** $c = 25(h + 12)$

3. ▦✎ **GRIDDED RESPONSE** The Music Shop records the number of CDs sold for 5 months. What fraction of CDs were sold in April? Write the fraction in simplest form.

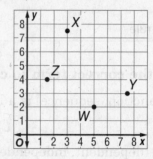

| Month | Number of CDs Sold |
|---|---|
| January | 50 |
| February | 35 |
| March | 42 |
| April | 110 |
| May | 98 |

4. Which point on the grid below corresponds to the ordered pair $\left(3, 7\frac{1}{2}\right)$?

   **A.** Point $W$     **C.** Point $Y$

   **B.** Point $X$     **D.** Point $Z$

5. ▦✎ **GRIDDED RESPONSE** The Geography Club is selling wrapping paper for a fundraiser. They ordered 16 cases. If each student in the club takes $\frac{2}{3}$ of a case, how many students are in the club?

6. Mrs. Washington has 25 students in her class. If each student needs 4 file cards, which equation can be used to find $s$, the total number of file cards?

   **F.** $s = 25 \div 4$     **H.** $s = 25 - 4$

   **G.** $s = 25 \times 4$     **I.** $s = 25 + 4$

7. The formula $V = \frac{1}{3}Bh$ can be used to find the volume of a pyramid.

   Which of the following **best** represents $\frac{1}{3}$?

   **A.** 0.33     **C.** 3

   **B.** 0.67     **D.** 3.3

8. **SHORT RESPONSE** There are 16 cars and 64 passengers scheduled to go to a concert. Write a ratio in simplest form that compares the number of passengers to the number of cars.

9. Mariel spends $0.75 every time she plays her favorite video game. She has $6.75 to spend. Which inequality shows how many times Mariel can play the video game?

    **F.** $0.75x < 6.75$

    **G.** $0.75x \leq 6.75$

    **H.** $0.75x > 6.75$

    **I.** $0.75x \geq 6.75$

10. **SHORT RESPONSE** Jacob multiplied 30 by 0.25 and got 7.5. Zane said that it was wrong because when you multiply two numbers, the product is always greater than both of the numbers. Explain who is correct.

11. Which of the following ordered pairs is located inside the graph of the square?

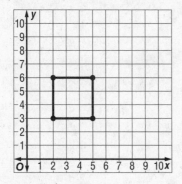

    **A.** (4, 0)

    **B.** (1, 6)

    **C.** (2, 2)

    **D.** (4, 5)

12. **SHORT RESPONSE** The table shows J.T.'s training schedule for a marathon.

| Day | Running Time (min) |
|-----|--------------------|
| 1   | 20                 |
| 2   | 22                 |
| 3   | 24                 |
| 4   | 26                 |

If the pattern continues, how many minutes will he run on Day 8?

13. **EXTENDED RESPONSE** The graph shows the cost of renting a car from Speedy Rental for various miles driven.

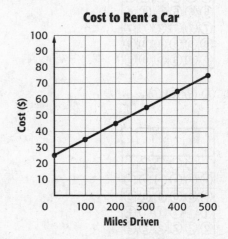

**Cost to Rent a Car**

*Part A* Make a function table for the graph.

*Part B* Write a function rule for the data.

*Part C* What is the cost in dollars of renting the car and driving 400 miles?

*Part D* Describe the relationship.

# Student Recording Sheet

*Use this recording sheet with the Standardized Test Practice pages.*

**Fill in the correct answer. For gridded-response questions, write your answers in the boxes on the answer grid and fill in the bubbles to match your answers.**

1. Ⓐ Ⓑ Ⓒ Ⓓ

2. Ⓕ Ⓖ Ⓗ Ⓘ

3.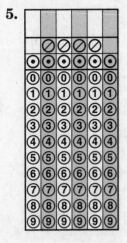

4. Ⓐ Ⓑ Ⓒ Ⓓ

5.

6. Ⓕ Ⓖ Ⓗ Ⓘ

7. Ⓐ Ⓑ Ⓒ Ⓓ

8. _____

9. Ⓕ Ⓖ Ⓗ Ⓘ

10. _____

11. Ⓐ Ⓑ Ⓒ Ⓓ

12. _____

## Extended Response
Record your answers for Exercise 13 on the back of this paper.

# Extended-Response Test

Demonstrate your knowledge by giving a clear, concise solution to each problem. Be sure to include all relevant drawings and justify your answers. You may show your solution in more than one way or investigate beyond the requirements of the problem. If necessary, record your answer on another piece of paper.

1. **a.** Write the rule for the function table at the right.

| Input ($x$) | Output (_____) |
|---|---|
| 1 | 2 |
| 2 | 5 |
| 3 | 8 |
| 4 | |
| 6 | |
| 10 | |

   **b.** Complete the function table at the right.

2. **TICKETS** The table below shows the cost of tickets to a county fair.

| Number of Tickets | 1 | 2 | 3 | 4 | 5 |
|---|---|---|---|---|---|
| Total Cost ($) | 7 | 14 | 21 | 28 | 35 |

   **a.** Use words to describe the rule to find the total cost of tickets to the fair as a function of the number of tickets purchased.

   **b.** Use symbols to describe the total cost of tickets to the fair as a function of the number of tickets purchased.

   **c.** Write an equation to represent the function displayed in the table. Define the variables.

   **d.** How much will it cost a family to buy 7 tickets?

   **e.** The sixth-grade class took a field trip to the fair and spent at least $420 on tickets. Write and solve an inequality to find the number of tickets they bought.

# Extended-Response Rubric

| Score | Description |
|:---:|:---|
| 4 | A score of four is a response in which the student demonstrates a thorough understanding of the mathematics concepts and/or procedures embodied in the task. The student has responded correctly to the task, used mathematically sound procedures, and provided clear and complete explanations and interpretations.<br><br>The response may contain minor flaws that do not detract from the demonstration of a thorough understanding. |
| 3 | A score of three is a response in which the student demonstrates an understanding of the mathematics concepts and/or procedures embodied in the task. The student's response to the task is essentially correct with the mathematical procedures used and the explanations and interpretations provided demonstrating an essential but less than thorough understanding.<br><br>The response may contain minor flaws that reflect inattentive execution of mathematical procedures or indications of some misunderstanding of the underlying mathematics concepts and/or procedures. |
| 2 | A score of two indicates that the student has demonstrated only a partial understanding of the mathematics concepts and/or procedures embodied in the task. Although the student may have used the correct approach to obtaining a solution or may have provided a correct solution, the student's work lacks an essential understanding of the underlying mathematical concepts.<br><br>The response contains errors related to misunderstanding important aspects of the task, misuse of mathematical procedures, or faulty interpretations of results. |
| 1 | A score of one indicates that the student has demonstrated a very limited understanding of the mathematics concepts and/or procedures embodied in the task. The student's response is incomplete and exhibits many flaws. Although the student's response has addressed some of the conditions of the task, the student reached an inadequate conclusion and/or provided reasoning that was faulty or incomplete.<br><br>The response exhibits many flaws or may be incomplete. |
| 0 | A score of zero indicates that the student has provided no response at all, or a completely incorrect or uninterpretable response, or demonstrated insufficient understanding of the mathematics concepts and/or procedures embodied in the task. For example, a student may provide some work that is mathematically correct, but the work does not demonstrate even a rudimentary understanding of the primary focus of the task. |

# Test, Form 1A

**Write the letter for the correct answer in the blank at the right of each question.**

1. Which of the following sets of values completes the function table?

| Input ($x$) | $4x + 2$ | Output ($y$) |
|:---:|:---:|:---:|
| 4 | $4(4) + 2$ | ▪ |
| 5 | $4(5) + 2$ | ▪ |
| 6 | $4(6) + 2$ | ▪ |

   **A.** 16, 20, 24          **C.** 18, 22, 26

   **B.** 18, 19, 20          **D.** 0, 1, 2             **1.** _____

2. Molly is buying packages of party favors for her birthday party. Using the table as a guide, how many packages will she need to buy to have 24 favors?

| Party Favors | |
|:---:|:---:|
| **Number of Packages** | **Number of Favors** |
| 3 | 6 |
| 6 | 12 |

   **F.** 8          **H.** 12

   **G.** 10        **I.** 14             **2.** _____

**Use the table below for Exercises 3 and 4.**

| Position | 1 | 2 | 3 | 4 | $n$ |
|:---|:---:|:---:|:---:|:---:|:---:|
| **Value of Term** | 2 | 4 | 6 | 8 | ▪ |

3. What is the rule to find the value of the missing term?

   **A.** $\frac{2}{n}$      **B.** $n + 2$      **C.** $2n$      **D.** $n - 2$         **3.** _____

4. What is the value of the twelfth term in the sequence?

   **F.** 10       **G.** 12       **H.** 16       **I.** 24         **4.** _____

5. The graph shows the total money earned at a fundraising car wash. Which equation can be used to find the total earned $y$ for the number of cars washed $x$?

   **A.** $y = x + 5$         **C.** $y = 5x$

   **B.** $y = \frac{x}{5}$          **D.** $y = x - 5$

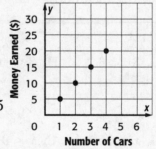

         **5.** _____

6. A pizzeria charges \$12 per pizza plus an addition \$3 for delivery. Which equation represents the cost of having any number of pizzas delivered?

   **F.** $c = 12p$    **G.** $c = 3p$    **H.** $c = 12p + 3$   **I.** $c = 12 + 3p$      **6.** _____

# Test, Form 1A    *(continued)*

**Use the following information for Exercises 7–9. Ryan earns $20 for every lawn that he mows.**

7. Which equation can be used to find $t$, the total amount Ryan will earn after mowing $n$ lawns?

   **A.** $t = 20n$    **B.** $n = 20t$    **C.** $t = 20 + n$    **D.** $n = 20 + t$    7. _____

8. How much will Ryan earn if he mows 15 lawns?

   **F.** $40    **G.** $150    **H.** $200    **I.** $300    8. _____

9. Which set of ordered pairs represents the relationship between the number of lawns Ryan mows and the money he earns?

   **A.** (20, 1), (40, 2), (60, 3)    **C.** (1, 20), (2, 40), (3, 60)

   **B.** (1, 20), (2, 30), (3, 40)    **D.** (0, 20), (1, 40), (2, 60)    9. _____

10. Which of the following is a solution of the inequality $h + 9 < 20$?

    **F.** 13    **G.** 12    **H.** 11    **I.** 10    10. _____

11. The inequality $a < 10$ represents the ages $a$ that qualify for a child ticket. Which children in the Rogers' family qualify for a child ticket?

    **A.** Chris, Megan, Piper

    **B.** Piper only

    **C.** Chris and Megan

    **D.** Piper and Mark

| Rogers' Family Ages | |
|---|---|
| Chris | 5 |
| Megan | 8 |
| Piper | 10 |
| Mark | 12 |

11. _____

12. Which inequality is graphed below?

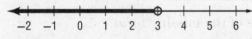

   **F.** $x \geq 3$    **G.** $x < 3$    **H.** $x \leq 3$    **I.** $x > 3$    12. _____

13. Miguel has at least $250 in his savings account. Which inequality represents this situation?

    **A.** $m < 250$    **B.** $m > 250$    **C.** $m \leq 250$    **D.** $m \geq 250$    13. _____

14. Which of the following inequalities has the solution shown below?

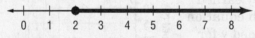

   **F.** $3x \leq 6$    **G.** $3x < 6$    **H.** $3x \geq 6$    **I.** $3x > 6$    14. _____

**Solve each inequality.**

15. $3 + x \geq 12$

   **A.** $x \geq 9$    **B.** $x \geq 15$    **C.** $x \leq 9$    **D.** $x \leq 15$    15. _____

16. $5x < 30$

   **F.** $x < 150$    **G.** $x < 6$    **H.** $x > 150$    **I.** $x > 6$    16. _____

17. $\frac{x}{3} \leq 6$

   **A.** $x \leq 2$    **B.** $x < 2$    **C.** $x \leq 18$    **D.** $x < 18$    17. _____

# Test, Form 1B

**Write the letter for the correct answer in the blank at the right of each question.**

1. Which of the following sets of values completes the function table?

| Input ($x$) | $3x - 1$ | Output ($y$) |
|---|---|---|
| 2 | $3(2) - 1$ | ■ |
| 3 | $3(3) - 1$ | ■ |
| 4 | $3(4) - 1$ | ■ |

   **A.** 0, 1, 2          **C.** 5, 6, 7

   **B.** 5, 8, 11        **D.** 6, 9, 12

   1. _____

2. Mrs. Miller is buying hot dog buns for a cookout. Using the table as a guide, how many packages will she need to buy to have 48 buns?

| Hot Dog Buns | |
|---|---|
| Number of Packages | Number of Buns |
| 2 | 16 |
| 4 | 32 |

   **F.** 6          **H.** 10

   **G.** 8          **I.** 12

   2. _____

**Use the table below for Exercises 3 and 4.**

| Position | 1 | 2 | 3 | 4 | $n$ |
|---|---|---|---|---|---|
| Value of Term | 3 | 6 | 9 | 12 | ■ |

3. What is the rule to find the value of the missing term?

   **A.** $\frac{3}{n}$          **B.** $3n$          **C.** $n + 2$          **D.** $n + 3$

   3. _____

4. What is the value of the twelfth term in the sequence?

   **F.** 9          **G.** 15          **H.** 24          **I.** 36

   4. _____

5. The graph shows the total cost of pizzas from a pizzeria. Which equation can be used to find the total cost $y$ for any number of pizzas $x$?

   **A.** $y = 10x$          **C.** $y = x - 10$

   **B.** $y = \frac{x}{10}$          **D.** $y = x + 10$

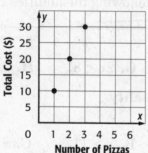

   5. _____

6. Admission to the county fair is $5. It costs an additional $0.50 for each ride ticket. Which equation represents the cost of going to the fair and buying any number of ride tickets?

   **F.** $c = 0.5t$          **H.** $c = 5t$

   **G.** $c = 0.5t + 5$          **I.** $c = 0.5 + 5t$

   6. _____

# Test, Form 1B   (continued)

**Use the following information for Exercises 7–9. Malia earns $5 for every hour that she babysits.**

7. Which equation can be used to find $t$, the total amount Malia will earn after babysitting $h$ hours?

   **A.** $h = 5 + t$   **B.** $t = 5 + h$   **C.** $h = 5t$   **D.** $t = 5h$

   7. _____

8. How much will Malia earn if she babysits for 8 hours?

   **F.** $10   **G.** $25   **H.** $40   **I.** $50

   8. _____

9. Which set of ordered pairs represents the relationship between the number of hours Malia babysits and the money she earns?

   **A.** (5, 1), (10, 2), (15, 3), (20, 4)   **C.** (1, 5), (2, 15), (3, 25), (4, 30)
   **B.** (1, 5), (2, 10), (3, 15), (4, 20)   **D.** (0, 5), (1, 10), (2, 15), (3, 20)

   9. _____

10. Which of the following is a solution of the inequality $y - 5 \geq 8$?

    **F.** 15   **G.** 12   **H.** 10   **I.** 8

    10. _____

11. The inequality $h \geq 48$ represents the minimum height $h$ necessary to ride a certain roller coaster. Who can ride the roller coaster?

    **A.** Sara only

    **B.** Anna only

    **C.** Anna and Sara

    **D.** Anna, Patrick, and Miguel

    | Heights (in.) | |
    | --- | --- |
    | Miguel | 42 |
    | Patrick | 45 |
    | Anna | 48 |
    | Sara | 52 |

    11. _____

12. Which inequality is graphed below?

    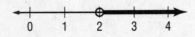

    **F.** $t \geq 2$   **G.** $t \leq 2$   **H.** $t > 2$   **I.** $t < 2$

    12. _____

13. Zachary can spend at most $100 on new clothes. Which inequality represents this situation?

    **A.** $s < 100$   **B.** $s > 100$   **C.** $s \leq 100$   **D.** $s \geq 100$

    13. _____

14. Which of the following inequalities has the solution shown below?

    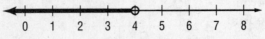

    **F.** $2x \leq 8$   **G.** $2x < 8$   **H.** $2x \geq 8$   **I.** $2x > 8$

    14. _____

**Solve each inequality.**

15. $6 + x \leq 17$

    **A.** $x \leq 11$   **B.** $x \geq 11$   **C.** $x \leq 23$   **D.** $x \leq 23$

    15. _____

16. $4x \geq 12$

    **F.** $x \geq 3$   **G.** $x \leq 12$   **H.** $x \geq 48$   **I.** $x \leq 48$

    16. _____

17. $\frac{x}{8} < 8$

    **A.** $x > 64$   **B.** $x < 64$   **C.** $x > 1$   **D.** $x < 1$

    17. _____

# Test, Form 2A

**Write the letter for the correct answer in the blank at the right of each question.**

1. Which of the following sets of values completes the function table?

| Input ($x$) | $5x - 4$ | Output ($y$) |
|---|---|---|
| 3 | $5(3) - 4$ |  |
| 11 | $5(11) - 4$ | |
| 19 | $5(19) - 4$ | |

   **A.** 15, 55, 95   **B.** 15, 21, 29   **C.** 11, 17, 25   **D.** 11, 51, 91

   1. _____

2. What is the rule to find the value of the missing term in the table?

| Position | 1 | 2 | 3 | 4 | $n$ |
|---|---|---|---|---|---|
| Value of Term | 5 | 10 | 15 | 20 |  |

   **F.** $n + 4$       **G.** $5n$       **H.** $n + 5$       **I.** $n \div 5$

   2. _____

3. The graph shows the charges for a movie rental club in a month. Which equation can be used to find the total charge $y$ for any number of rentals $x$ in a month?

   **A.** $y = 2x$                **C.** $y = 2x + 3$

   **B.** $y = 3x$                **D.** $y = 3x + 2$

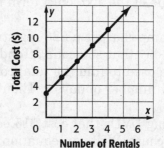

   3. _____

4. Which inequality is graphed below?

   **F.** $x \geq 15$       **G.** $x < 15$       **H.** $x \leq 15$       **I.** $x > 15$

   4. _____

5. Which of the following is a solution of the inequality $m - 12 < 20$?
   **A.** 34       **B.** 33       **C.** 32       **D.** 31

   5. _____

6. Which of the following inequalities has the solution shown below?

   **F.** $6n > 24$     **G.** $6n \geq 24$     **H.** $6n \leq 24$     **I.** $6n < 24$

   6. _____

**Solve each inequality.**

7. $c - 7 \leq 10$
   **A.** $c \leq 3$       **B.** $c \geq 17$       **C.** $c \geq 3$       **D.** $c \leq 17$

   7. _____

8. $4b > 20$
   **F.** $b > 5$       **G.** $b < 5$       **H.** $b > 80$       **I.** $b < 80$

   8. _____

9. $\dfrac{d}{4} < 16$
   **A.** $d > 4$       **B.** $d < 4$       **C.** $d < 64$       **D.** $d > 64$

   9. _____

# Test, Form 2A    (continued)

**For Exercises 10–12, find the rule for each function table.**

10.

| Input (x) | Output (y) |
|-----------|------------|
| 1 | 3 |
| 2 | 4 |
| 4 | 6 |

11.

| Input (x) | Output (y) |
|-----------|------------|
| 0 | 0 |
| 3 | 9 |
| 6 | 18 |

12.

| Input (x) | Output (y) |
|-----------|------------|
| 4 | 1 |
| 8 | 2 |
| 12 | 3 |

10. _____

11. _____

12. _____

**Use the table below for Exercises 13 and 14.**

| Position | 9 | 10 | 11 | 12 | n |
|----------|---|----|----|----|---|
| Value of Term | 3 | 4 | 5 | 6 | ▪ |

13. Use words and symbols to describe the value of each term as a function of its position.

13. _____

14. Find the value of the sixteenth term in the sequence.

14. _____

15. A summer camp charges a $25 registration fee plus an additional $10 for each day that someone attends the camp. Write an equation that could be used to find the total cost y for someone to attend the camp for any number of days x. Then graph the equation.

15.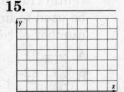

16. Hugo is buying DVDs that cost $15 each. He has a coupon for $5 off his total purchase. Write an equation to find c the total amount he will spend on any number of DVDs d. Then use the equation to find the amount he will spend if he buys 6 DVDs.

16. _____

**Write an equation to represent the function.**

17.

| Input, x | 1 | 2 | 3 | 4 |
|----------|---|---|---|---|
| Output, y | 4 | 8 | 12 | 16 |

18.

| Input, x | 1 | 2 | 3 | 4 |
|----------|---|---|---|---|
| Output, y | 2 | 5 | 8 | 11 |

17. _____

18. _____

19. Is 12, 13, or 14 a solution of the inequality $4x > 52$?

19. _____

20. Write an inequality to represent the statement *band practice will be no longer than 45 minutes*. Then graph the inequality on a number line.

20. _____

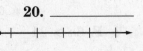

**Solve each inequality. Graph the solution on a number line.**

21. $x - 9 < 14$

21. _____

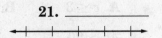

22. $\frac{n}{6} \geq 3$

22. _____

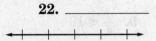

# Test, Form 2B

**Write the letter for the correct answer in the blank at the right of each question.**

**1.** Which of the following sets of values completes the function table?

| Input ($x$) | $2x + 6$ | Output ($y$) |
|:---:|:---:|:---:|
| 3 | $2(3) + 6$ | ▪ |
| 9 | $2(9) + 6$ | ▪ |
| 17 | $2(17) + 6$ | ▪ |

    **A.** 6, 18, 34    **B.** 12, 24, 40    **C.** 12, 18, 26    **D.** 0, 12, 28

**1.** _____

**2.** What is the rule to find the value of the missing term in the table?

| Position | 1 | 2 | 3 | 4 | $n$ |
|:---|:---:|:---:|:---:|:---:|:---:|
| Value of Term | 5 | 6 | 7 | 8 | ▪ |

    **F.** $n + 1$    **G.** $n + 4$    **H.** $5n$    **I.** $4n + 1$

**2.** _____

**3.** The graph shows the total cost of a zoo membership for a family. Which equation can be used to find the total cost $y$ for any number of family members $x$?

    **A.** $y = 10x$         **C.** $y = 10x + 25$

    **B.** $y = 25x$         **D.** $y = 25x + 10$

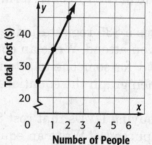

**3.** _____

**4.** Which inequality is graphed below?

    **F.** $r \leq 13$    **G.** $r < 13$    **H.** $r \geq 13$    **I.** $r > 13$

**4.** _____

**5.** Which of the following is a solution of the inequality $3x \geq 15$?

    **A.** 0        **B.** 2        **C.** 4        **D.** 6

**5.** _____

**6.** Which of the following inequalities has the solution shown below?

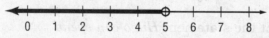

    **F.** $4n \geq 20$    **G.** $4n \leq 20$    **H.** $4n > 20$    **I.** $4n < 20$

**6.** _____

**Solve each inequality.**

**7.** $x - 3 \leq 7$

    **A.** $x \leq 4$    **B.** $x \geq 4$    **C.** $x \geq 10$    **D.** $x \leq 10$

**7.** _____

**8.** $3b < 18$

    **F.** $b < 6$    **G.** $b > 6$    **H.** $b > 54$    **I.** $b < 54$

**8.** _____

**9.** $\frac{y}{3} > 9$

    **A.** $y > 3$    **B.** $y < 3$    **C.** $y > 27$    **D.** $y < 27$

**9.** _____

# Test, Form 2B  (continued)

**For Exercises 10–12, find the rule for each function table.**

10.

| Input (x) | Output (y) |
|-----------|------------|
| 1 | 4 |
| 2 | 8 |
| 4 | 16 |

11.

| Input (x) | Output (y) |
|-----------|------------|
| 0 | 0 |
| 3 | 1 |
| 9 | 3 |

12.

| Input (x) | Output (y) |
|-----------|------------|
| 3 | 1 |
| 5 | 3 |
| 8 | 6 |

10. _____

11. _____

12. _____

**Use the table below for Exercises 13 and 14.**

| Position | 1 | 2 | 3 | 4 | n |
|----------|---|---|---|---|---|
| Value of Term | 8 | 16 | 24 | 32 | ■ |

13. Use words and symbols to describe the value of each term as a function of its position.

13. _____

14. Find the value of the fifteenth term in the sequence.

14. _____

15. A gym charges a $35 registration fee plus an additional $20 for each month that you attend. Write an equation that could be used to find the total cost $y$ for someone to attend the gym for any number of months $x$. Then graph the equation.

15.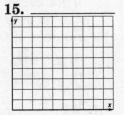

16. Lauretta is buying DVDs that cost $9 each. She has a coupon for $6 off her total purchase. Write an equation to find $c$ the total amount she will spend on any number of DVDs $d$. Then use the equation to find the amount she will spend if she buys 8 DVDs.

16. _____

**Write an equation to represent the function.**

17.

| Input, x | 1 | 2 | 3 | 4 |
|----------|---|---|---|---|
| Output, y | 3 | 6 | 9 | 12 |

18.

| Input, x | 1 | 2 | 3 | 4 |
|----------|---|---|---|---|
| Output, y | 2 | 6 | 10 | 14 |

17. _____

18. _____

19. Is 11, 12, or 13 a solution of the inequality $3x < 36$?

19. _____

20. Write an inequality to represent the statement *Hugo can spend no more than $10 on lunch*. Then graph the inequality on a number line.

20. _____

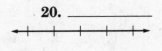

**Solve each inequality. Graph the solution on a number line.**

21. $x - 4 \geq 12$

21. _____

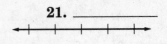

22. $\frac{x}{12} < 2$

22. _____

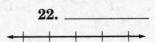

# Test, Form 3A

**Write the correct answer in the blank at the right of each question.**

1. Which of the numbers 4, 5, or 6 is a solution of $x + 5 > 10$?

1. _____

2. Complete the table.

| Input ($x$) | $3x - 2$ | Output ($y$) |
|---|---|---|
| 3 | ▪ | ▪ |
| 6 | ▪ | ▪ |
| ▪ | ▪ | 22 |

| Input ($x$) | $3x - 2$ | Output ($y$) |
|---|---|---|
| 3 | | |
| 6 | | |
| | | 22 |

2. _____

**Find the rule for each function table.**

3.

| Input ($x$) | Output ($y$) |
|---|---|
| 0 | 0 |
| 2 | 6 |
| 5 | 15 |

4.

| Input ($x$) | Output ($y$) |
|---|---|
| 4 | 1 |
| 12 | 3 |
| 20 | 5 |

5.

| Input ($x$) | Output ($y$) |
|---|---|
| 0 | 3 |
| 1 | 5 |
| 2 | 7 |

3. _____

4. _____

5. _____

**Write an inequality to represent each situation. Then graph the inequality.**

6. Anyone over age fourteen must buy an adult ticket.

6. _____

    ⊢—+—+—+—+—▸

7. Kendra can spent at most $8 on lunch.

7. _____

    ⊢—+—+—+—+—▸

**Solve each inequality. Then graph the solution on a number line.**

8. $a + 14 < 20$

8. _____

    ⊢—+—+—+—+—▸

9. $c - 15 \geq 7$

9. _____

    ⊢—+—+—+—+—▸

10. $8m \leq 48$

10. _____

    ⊢—+—+—+—+—▸

11. $\dfrac{h}{5} > 9$

11. _____

    ⊢—+—+—+—+—▸

12. Riaz wants to spend no more than $1,250 on his trip. The airfare is $439. Write and solve an inequality to find the maximum amount he can spend on the rest of his trip.

12. _____

**Course 1 • Chapter 8** Functions and Inequalities

# Test, Form 3A   (continued)

**Use the table below for Exercises 13 and 14.**

| Position | 1 | 2 | 3 | 4 | $n$ |
|---|---|---|---|---|---|
| Value of Term | 9 | 18 | 27 | 36 | ▪ |

13. Use words and symbols to describe the value of each term as a function of its position.

13. _____

14. Find the value of the sixteenth term in the sequence.

14. _____

**Write an equation to represent the function.**

15.

| Input, $x$ | 1 | 2 | 3 | 4 |
|---|---|---|---|---|
| Output, $y$ | 5 | 10 | 15 | 20 |

15. _____

16.

| Input, $x$ | 1 | 2 | 3 | 4 |
|---|---|---|---|---|
| Output, $y$ | 9 | 16 | 23 | 30 |

16. _____

**Use the following information for Exercises 17–20. An adult male gorilla eats about 40 pounds of vegetation every day.**

17. Write an equation to find $p$, the number of pounds of vegetation an adult male gorilla eats in $d$ days.

17. _____

18. Make a table to show the relationship between the number of pounds $p$ an adult male gorilla eats in $d$ days.

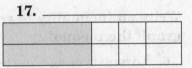

18.

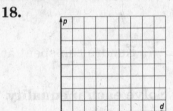

19. Graph the ordered pairs from the table in Exercise 18.

19. _____

20. Analyze the graph in Exercise 19.

20. _____

**Graph each equation.**

21. $y = 0.5x + 2$

21.

22. $y = x + 1$

22.

# Test, Form 3B

**Write the correct answer in the blank at the right of each question.**

1. Which of the numbers 4, 5, or 6 is a solution of $x + 7 > 12$?

1. _____

2. Complete the table.

| Input ($x$) | $2x + 4$ | Output ($y$) |
|---|---|---|
| 2 | ■ | ■ |
| 5 | ■ | ■ |
| ■ | ■ | 24 |

| Input ($x$) | $2x + 4$ | Output ($y$) |
|---|---|---|
| 2 | | |
| 5 | | |
| | | 24 |

2. _____

**Find the rule for each function table.**

3.

| Input ($x$) | Output ($y$) |
|---|---|
| 3 | 1 |
| 12 | 4 |
| 18 | 6 |

4.

| Input ($x$) | Output ($y$) |
|---|---|
| 5 | 14 |
| 7 | 20 |
| 11 | 32 |

5.

| Input ($x$) | Output ($y$) |
|---|---|
| 0 | 0 |
| 3 | 6 |
| 5 | 10 |

3. _____

4. _____

5. _____

**Write an inequality to represent each situation. Then graph the inequality.**

6. The decorations cost more than $12.

6. _____

7. Children ages 6 and under qualify for free admission.

7. _____

**Solve each inequality. Then graph the solution on a number line.**

8. $16 < m - 17$

8. _____

9. $d + 12 \leq 22$

9. _____

10. $5x \geq 75$

10. _____

11. $\frac{n}{3} < 12$

11. _____

12. Julian wants to spend no more than $1,250 on his trip. The airfare is $358. Write and solve an inequality to find the maximum amount he can spend on the rest of his trip.

12. _____

# Test, Form 3B    (continued)

**Use the table below for Exercises 13 and 14.**

| Position | 1 | 2 | 3 | 4 | $n$ |
|---|---|---|---|---|---|
| Value of Term | 8 | 16 | 24 | 32 | ■ |

**13.** Use words and symbols to describe the value of each term as a function of its position.

13. _____

**14.** Find the value of the sixteenth term in the sequence.

14. _____

**Write an equation to represent the function.**

**15.**

| Input, $x$ | 1 | 2 | 3 | 4 |
|---|---|---|---|---|
| Output, $y$ | 6 | 11 | 16 | 21 |

15. _____

**16.**

| Input, $x$ | 1 | 2 | 3 | 4 |
|---|---|---|---|---|
| Output, $y$ | 1 | 4 | 7 | 10 |

16. _____

**Use the following information for Exercises 17–20. A giraffe can eat about 75 pounds of vegetation every day.**

**17.** Write an equation to find $p$, the number of pounds of vegetation a giraffe eats in $d$ days.

17. _____

**18.** Make a table to show the relationship between the number of pounds $p$ a giraffe eats in $d$ days.

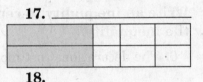

18.

**19.** Graph the ordered pairs from the table in Exercise 18.

19. _____

**20.** Analyze the graph in Exercise 19.

20. _____

**Graph each equation.**

**21.** $y = \dfrac{1}{4}x + 1$

21.

**22.** $y = 2x$

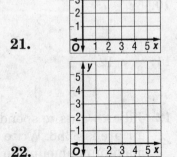

# Are You Ready?

## Review

### Example

**Find $\frac{3}{4} \times 12$.**

$$\frac{3}{4} \times 12 = \frac{3}{4} \times \frac{12}{1} \qquad \text{Write 12 as } \frac{12}{1}.$$

$$= \frac{3 \times \overset{3}{\cancel{12}}}{\underset{1}{\cancel{4}} \times 1} \qquad \text{Divide the numerator and denominator by 4.}$$

$$= \frac{9}{1} \text{ or } 9 \qquad \text{Simplify.}$$

### Exercises

**Evaluate each expression.**

1. $\frac{1}{5} \times 5$

2. $\frac{1}{10} \times 20$

3. $\frac{1}{3} \times 21$

4. $\frac{1}{4} \times 48$

5. Find the area of the rectangle.

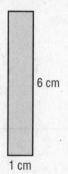

6 cm

1 cm

6. **PAPER** The size of a typical page from Paige's notebook is 11 inches long by 8 inches wide. What is the area of a typical page from her notebook?

1. _____

2. _____

3. _____

4. _____

5. _____

6. _____

# Are You Ready?

## Practice

**Evaluate each expression.**

1. $\frac{1}{2} \times 2$

2. $\frac{2}{5} \times 10$

3. $\frac{7}{10} \times 10$

4. $\frac{3}{4} \times 8$

5. $\frac{2}{3} \times 33$

6. $\frac{1}{4} \times 44$

7. $\frac{5}{6} \times 36$

1. _____

2. _____

3. _____

4. _____

5. _____

6. _____

7. _____

**Find the area of each rectangle.**

8.

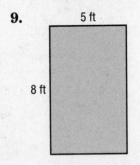
7 cm
3 cm

9.
5 ft

8 ft

8. _____

9. _____

10. **SANDBOX** A sandbox at a playground is
    3 yards long by 2 yards wide. What is the area
    of the sandbox?

10. _____

# Are You Ready?

## Apply

1. **FARMING** A farmer has a field that is 40 yards long by 20 yards wide. What is the total area of the field?

2. **WALLPAPER** Sydney is going to wallpaper her bedroom wall. If her wall is 10 feet long by 8 feet tall, how much wallpaper will she need?

3. **GARDENING** Elena wants to put a new layer of soil in her garden. She needs to know the area of her garden so that she knows how much soil to buy. If the garden is 14 meters by 6 meters, how much area does she need to cover?

4. **TILES** Mr. McCabe is buying tiles for his kitchen floor. The floor is 12 feet long by 11 feet wide. If each tile is 1 square foot, how many tiles does he need?

5. **STORM** A weather station reported that a strong storm covered an area of land that was 13 kilometers long by 15 kilometers wide. How much land area was covered by the storm?

6. **WILDFIRES** A spokesperson from the forest fire service said that a wildfire caused a patch of forest that was 20 miles long by 20 miles wide to be burned. How many square miles of forest were burned by the wildfire?

# Diagnostic Test

**Evaluate each expression.**

1. $\frac{2}{3} \times 3$

2. $\frac{1}{2} \times 30$

3. $4 \times \frac{3}{4}$

4. $\frac{1}{2} \times 24$

5. $\frac{3}{5} \times 15$

6. $\frac{2}{3} \times 18$

7. $\frac{4}{5} \times 50$

1. _____

2. _____

3. _____

4. _____

5. _____

6. _____

7. _____

**Find the area of each rectangle.**

8.
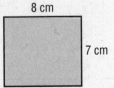
8 cm
7 cm

9.
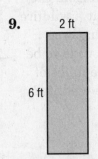
2 ft
6 ft

8. _____

9. _____

10. **POOL COVER** A pool cover is 12 yards long by 12 yards wide. What is the area of the pool cover?

10. _____

# Pretest

**Find the area of each figure.**

**1.**

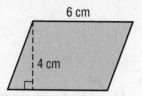

6 cm

4 cm

**2.**

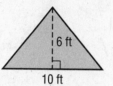

6 ft

10 ft

**3.**

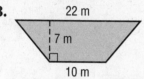

22 m

7 m

10 m

**4.**

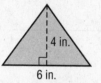

4 in.

6 in.

**5.**

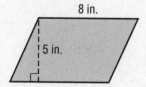

8 in.

5 in.

**6.**

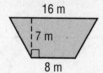

16 m

7 m

8 m

1. _____

2. _____

3. _____

4. _____

5. _____

6. _____

**7.** Find the area of the figure shown below.

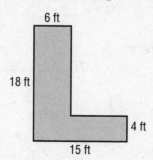

6 ft

18 ft

4 ft

15 ft

7. _____

# Chapter Quiz

**Find the area of each figure.**

**1.**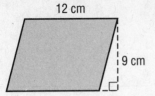
12 cm

9 cm

**2.**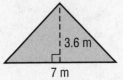
3.6 m

7 m

**3.**
11 ft

5 ft

4 ft

**4.**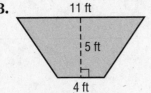
8 in.

15 in.

**5.**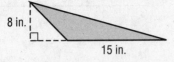
4 in.

8 in.

**6.** Find the height of a parallelogram with an area of 224 square meters and a base of 16 meters.

**7.** Find the height of a triangle with an area of 245 square inches and a base of 14 inches.

**8.** Find the height of a parallelogram with an area of 300 square yards and a base of 15 yards.

1. _____

2. _____

3. _____

4. _____

5. _____

6. _____

7. _____

8. _____

# Vocabulary Test

| | | |
|---|---|---|
| base | formula | polygon |
| composite figure | height | rhombus |
| congruent | parallelogram | |

**Fill in the blank to complete each sentence.**

1. A _____ is made up of triangles, quadrilaterals, semicircles, and other two-dimensional figures.

1. _____

2. _____ is a closed figure formed by 3 or more straight lines.

2. _____

3. Any side of a parallelogram is called its _____.

3. _____

4. The formula for the area of a _____ is $A = bh$.

4. _____

5. An equation that shows a relationship among certain quantities is called a(n) _____.

5. _____

6. The _____ of a parallelogram is the shortest distance between the parallel sides.

6. _____

**In your own words, define the terms.**

7. rhombus

7. _____

8. congruent

8. _____

# Standardized Test Practice

**Read each question. Then fill in the correct answer on the answer sheet provided by your teacher or on a sheet of paper.**

1. The table below shows the areas of a triangle where the height of the triangle stays the same, but the base changes.

| Area of Triangles | | |
|---|---|---|
| Height (units) | Base (units) | Area (square units) |
| 4 | 3 | 6 |
| 4 | 4 | 8 |
| 4 | 5 | 10 |
| 4 | 6 | 12 |
| 4 | $n$ | ▪ |

Which expression can be used to find the area of a triangle that has a height of 4 units and a base of $n$ units?

**A.** $\frac{n}{4}$

**B.** $\frac{4n}{2}$

**C.** $\frac{4}{2n}$

**D.** $4n$

2. ✎ **GRIDDED RESPONSE** José used a square baking pan to make a cake. The length of each side of the pan was 16 inches. Find the area of the pan in square inches.

3. Janet has a garden in the shape of parallelogram in her front yard. What is the area of the garden if it has a base of 10 feet of a height of 4 feet?

**F.** 20 ft²

**G.** 30 ft²

**H.** 40 ft²

**I.** 50 ft²

4. In the spreadsheet below, a formula applied to the values in columns $A$ and $B$ results in the values in column $C$. What is the formula?

| | A | B | C |
|---|---|---|---|
| 1 | 4 | 0 | 4 |
| 2 | 5 | 1 | 3 |
| 3 | 6 | 2 | 2 |
| 4 | 7 | 3 | 1 |

**A.** $C = A - B$     **C.** $C = A + B$

**B.** $C = A - 2B$    **D.** $C = A + 2B$

5. ▪ **SHORT RESPONSE** In Mrs. Tucker's classroom library, there are 168 fiction and 224 nonfiction books. What is the ratio of fiction to nonfiction books in simplest form?

6. Which expression gives the area of a triangle with a base of 8 units and height 3 units?

**F.** $8 \times 3$

**G.** $\frac{1}{2}(8 \times 3)$

**H.** $\frac{1}{2}(8 + 3)$

**I.** $(8 + 3) + (8 + 3)$

7. Ted is making three picture frames like the one shown below. What length of wood does Ted need for all three picture frames?

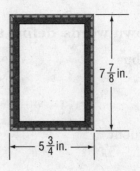

$7\frac{7}{8}$ in.

$5\frac{3}{4}$ in.

**A.** $11\frac{1}{2}$ in.     **C.** $27\frac{1}{4}$ in.

**B.** $15\frac{3}{4}$ in.     **D.** $81\frac{3}{4}$ in.

8. **THINK SOLVE EXPLAIN** **SHORT RESPONSE** Lynette is painting a 15-foot by 10-foot rectangular wall that has a 9-foot by 5-foot rectangular window at its center.

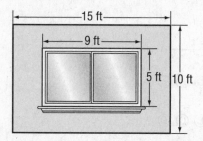

How many square feet of wall will she paint?

9. The cost of renting a car is shown in the advertisement.

Rental: $50 plus $0.10 per mile

Which of the following equations can be used to find $t$, the cost in dollars of the rental for $m$ miles?

**F.** $t = 0.10m + 25$

**G.** $t = 50 + 0.10$

**H.** $t = 50(m + 0.10)$

**I.** $t = 50 + 0.10m$

10. The area of a triangle is 30 square inches. What is the length of the base if the height is 6 centimeters?

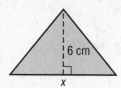

**A.** 12 cm

**B.** 10 cm

**C.** 5 cm

**D.** 3 cm

11. **GRIDDED RESPONSE** The road sign shows the distances from the highway exit to certain businesses.

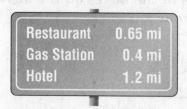

| Restaurant | 0.65 mi |
| Gas Station | 0.4 mi |
| Hotel | 1.2 mi |

What fraction of a mile is the restaurant from the exit?

12. For every $5 Marta earns mowing lawns, she puts $2 in her savings account. How much money will she have to earn in order to deposit $30 into her savings account?

**F.** $6

**G.** $12

**H.** $15

**I.** $75

13. **THINK SOLVE EXPLAIN** **EXTENDED RESPONSE** Ryan is painting a mural for his college art final. The mural is shaped like the figure shown below.

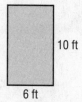

10 ft

6 ft

*Part A* Find the perimeter of the figure.

*Part B* Suppose Ryan doubles the side length of each side, what happens to the perimeter of the figure? Explain your reasoning.

# Student Recording Sheet

*Use this recording sheet with the Standardized Test Practice pages.*

**Fill in the correct answer. For gridded-response questions, write your answers in the boxes on the answer grid and fill in the bubbles to match your answers.**

1. Ⓐ Ⓑ Ⓒ Ⓓ

2.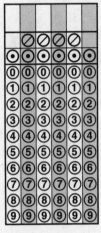

3. Ⓕ Ⓖ Ⓗ Ⓘ

4. Ⓐ Ⓑ Ⓒ Ⓓ

5. _____

6. Ⓕ Ⓖ Ⓗ Ⓘ

7. Ⓐ Ⓑ Ⓒ Ⓓ

8. _____

9. Ⓕ Ⓖ Ⓗ Ⓘ

10. Ⓐ Ⓑ Ⓒ Ⓓ

11.

12. Ⓕ Ⓖ Ⓗ Ⓘ

## Extended Response
Record your answers for Exercise 13 on the back of this paper.

# Extended-Response Test

**Demonstrate your knowledge by giving a clear, concise solution to each problem. Be sure to include all relevant drawings and justify your answers. You may show your solution in more than one way or investigate beyond the requirements of the problem. If necessary, record your answer on another piece of paper.**

1. Ms. Sharp is collecting fabric scraps to make a patchwork quilt. She is cutting quadrilaterals and triangles. For each figure, find out how much fabric the shape would use. Show how you found each answer. Use symbols and explain what the symbols mean.

   a.
   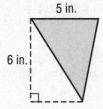

   b.
   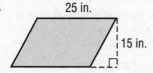

   c.

2. Show how to find the perimeter and area of the figure shown below.

   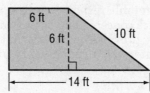

3. Refer to Exercise 2. Suppose you double the length of each measure in the figure. What happens to the perimeter of the figure? Explain your reasoning.

# Extended-Response Rubric

| Score | Description |
|:---:|:---|
| 4 | A score of four is a response in which the student demonstrates a thorough understanding of the mathematics concepts and/or procedures embodied in the task. The student has responded correctly to the task, used mathematically sound procedures, and provided clear and complete explanations and interpretations.<br><br>The response may contain minor flaws that do not detract from the demonstration of a thorough understanding. |
| 3 | A score of three is a response in which the student demonstrates an understanding of the mathematics concepts and/or procedures embodied in the task. The student's response to the task is essentially correct with the mathematical procedures used and the explanations and interpretations provided demonstrating an essential but less than thorough understanding.<br><br>The response may contain minor flaws that reflect inattentive execution of mathematical procedures or indications of some misunderstanding of the underlying mathematics concepts and/or procedures. |
| 2 | A score of two indicates that the student has demonstrated only a partial understanding of the mathematics concepts and/or procedures embodied in the task. Although the student may have used the correct approach to obtaining a solution or may have provided a correct solution, the student's work lacks an essential understanding of the underlying mathematical concepts.<br><br>The response contains errors related to misunderstanding important aspects of the task, misuse of mathematical procedures, or faulty interpretations of results. |
| 1 | A score of one indicates that the student has demonstrated a very limited understanding of the mathematics concepts and/or procedures embodied in the task. The student's response is incomplete and exhibits many flaws. Although the student's response has addressed some of the conditions of the task, the student reached an inadequate conclusion and/or provided reasoning that was faulty or incomplete.<br><br>The response exhibits many flaws or may be incomplete. |
| 0 | A score of zero indicates that the student has provided no response at all, or a completely incorrect or uninterpretable response, or demonstrated insufficient understanding of the mathematics concepts and/or procedures embodied in the task. For example, a student may provide some work that is mathematically correct, but the work does not demonstrate even a rudimentary understanding of the primary focus of the task. |

# Test, Form 1A

**Write the letter for the correct answer in the blank at the right of each question.**

1. An artist is using tiles in the shape of a parallelogram to make a mosaic. The tiles have the dimensions shown. What is the area of one tile?

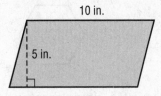

   **A.** 2 in²         **C.** 25 in²

   **B.** 15 in²       **D.** 50 in²

   1. _____

2. What is the height of a parallelogram with base 7 meters and an area of 84 square meters?

   **F.** 7 m          **H.** 294 m

   **G.** 12 m        **I.** 588 m

   2. _____

3. What is the area of the triangle?

   **A.** 24 in²

   **B.** 12 in²

   **C.** 6 in²

   **D.** 3.5 in²

   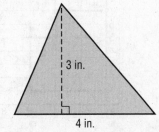

   3. _____

4. A triangle is cut from a piece of fabric. The triangle has a height of 8 inches and an area of 120 square inches. What is the length of the base of the triangle?

   **F.** 960 in.       **H.** 30 in.

   **G.** 480 in.      **I.** 15 in.

   4. _____

5. An eraser has the dimensions shown at the right. What is the area of the eraser?

   **A.** 19 cm²

   **B.** 39 cm²

   **C.** 78 cm²

   **D.** 120 cm²

   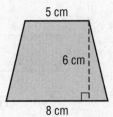

   5. _____

6. Which of the following is *not* a characteristic of the figure with vertices at coordinates $A(1, 1)$, $B(1, 2)$, $C(2, 2)$, and $D(2, 1)$?

   **F.** four right angles

   **G.** four vertices

   **H.** two sets of parallel lines

   **I.** two acute angles

   6. _____

**Course 1 • Chapter 9** Area

# Test, Form 1A    *(continued)*

**7.** The regular hexagon shown is enlarged so that its sides are 5 times as large. What affect does this have on the perimeter?

6 cm

A. The perimeter is 2.5 times greater.
B. The perimeter is 5 times greater.
C. The perimeter is the same.
D. The perimeter is 10 times greater.

7. _____

**8.** The blueprints for a new garden are shown. Each grid square on the blueprint has a length of 10 feet. What is the total distance, in feet, around the garden?

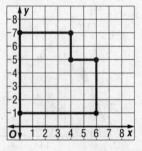

F. 22 feet                H. 220 feet
G. 24 feet                I. 240 feet

8. _____

**9.** A rectangle has vertices $A(1, 2)$, $B(1, 7)$, $C(4, 7)$, and $D(4, 2)$. What is the perimeter of the rectangle?

A. 3 units                C. 15 units
B. 5 units                D. 16 units

9. _____

**10.** What is the area of the figure shown below?

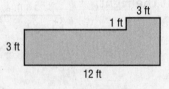

3 ft
1 ft
3 ft
12 ft

F. 39 ft²                H. 50 ft²
G. 41 ft²                I. 60 ft²

10. _____

**11.** Two acute triangles have perimeters of 20 meters and 100 meters respectively. How many times greater is the perimeter of the larger triangle?

A. 8 times                C. 5 times
B. 6 times                D. 4 times

11. _____

# Test, Form 1B

**Write the letter for the correct answer in the blank at the right of each question.**

1. A decorative rug in an office building has the dimensions shown. What is the area of the rug?
   A. 72 m²
   B. 36 m²
   C. 18 m²
   D. 6 m²

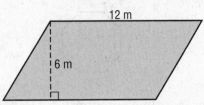

1. _____

2. What is the height of a parallelogram with base 6 meters and an area of 90 square meters?
   F. 540 m          H. 30 m
   G. 84 m           I. 15 m

2. _____

3. What is the area of the triangle?
   A. 24 in²
   B. 12 in²
   C. 6 in²
   D. 5.5 in²

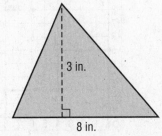

3. _____

4. Asia is designing a triangular-shaped window with a height of 15 inches and an area of 135 square inches. What is the length of the base of the window?
   F. 9 in.          H. 120 in.
   G. 18 in.         I. 2,025 in.

4. _____

5. A charm on Lucy's bracelet has the dimensions shown. What is the area of the charm?
   A. 270 mm²
   B. 96 mm²
   C. 48 mm²
   D. 24 mm²

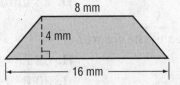

5. _____

6. Which of the following is *not* a characteristic of the figure with vertices at coordinates A(1, 3), B(1, 5), C(8, 5), and D(8, 3)?
   F. two sets of parallel lines
   G. four right angles
   H. four vertices
   I. four equal side lengths

6. _____

# Test, Form 1B   *(continued)*

7. The regular octagon shown is enlarged so that its sides are 3 times as large. What affect does this have on the perimeter?

5 in.

   A. The perimeter is 2 times greater.
   B. The perimeter is 3 times greater.
   C. The perimeter stays the same.
   D. The perimeter is 9 times greater.

7. _____

8. The blueprints for a new garden are shown. Each grid square on the blueprint has a length of 20 feet. What is the total distance around the garden?

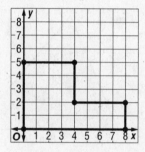

   F. 26 feet              H. 260 feet
   G. 52 feet              I. 520 feet

8. _____

9. A rectangle has vertices $A(0, 0)$, $B(0, 9)$, $C(4, 9)$, and $D(4, 0)$. What is the perimeter of the rectangle?

   A. 4 units             C. 13 units
   B. 9 units             D. 26 units

9. _____

10. What is the area of the figure?

   F. 66 ft²             H. 32 ft²
   G. 36 ft²             I. 30 ft²

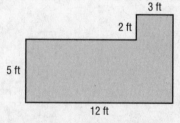

10. _____

11. Two squares have perimeters of 25 yards and 100 yards respectively. How many times greater is the perimeter of the larger square?

   A. 75 times           C. 4 times
   B. 5 times            D. 2 times

11. _____

# Test, Form 2A

**Write the letter for the correct answer in the blank at the right of each question.**

1. A package of sticky notes is in the shape of a parallelogram. The dimensions of one sticky note are shown. What is the area of one sticky note?

   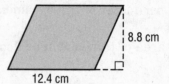

   **A.** 21.2 cm²          **C.** 54.56 cm²

   **B.** 42.4 cm²          **D.** 109.12 cm²          1. _____

2. What is the base of a parallelogram with height 5.6 meters and an area of 39.2 square meters?

   **F.** 2.8 m          **H.** 19.6 m

   **G.** 7 m          **I.** 33.6 m          2. _____

3. What is the area of the triangle?

   **A.** 180 m²

   **B.** 225 m²

   **C.** 360 m²

   **D.** 450 m²

   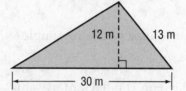

   3. _____

4. Antar is flying a triangular-shaped kite. It has a height of $4\frac{1}{2}$ feet and a base of $5\frac{3}{4}$ feet. What is the area of Antar's kite?

   **F.** $5\frac{1}{8}$ ft²          **H.** $12\frac{15}{16}$ ft²

   **G.** $10\frac{1}{4}$ ft²          **I.** $25\frac{7}{8}$ ft²          4. _____

5. A triangle has a base of 17 inches and an area of 127.5 square inches. What is the height of the triangle?

   **A.** 3.75 in.          **C.** 15 in.

   **B.** 7.5 in.          **D.** 30 in.          5. _____

6. What is the area of the trapezoid?

   **F.** 275 m²          **H.** 550 m²

   **G.** 330 m²          **I.** 660 m²          6. _____

   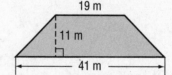

7. A farmer is installing a new fence. The coordinates of the vertices of the fence are $A(2, 2)$, $B(2, 6)$, $C(8, 6)$, and $D(8, 2)$. If each grid square has a length of 9 yards, how much wire is needed for the fence?

   **A.** 20 yards          **C.** 180 yards

   **B.** 24 yards          **D.** 216 yards          7. _____

# Test, Form 2A   (continued)

**For Exercises 8 and 9, refer to the figure at the right that shows the dimensions of Ramiro's basement floor.**

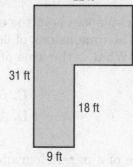

8. What is the perimeter of the basement floor?

8. _____

9. What is the area of the basement floor?

9. _____

**For Exercises 10 and 11, use the following information.**
**A rectangle has vertices $A(1, 3)$, $B(1, 10)$, $C(5, 10)$, and $D(5, 3)$.**

10. What is the length of each side of the rectangle?

10. _____

11. What is the perimeter of the rectangle?

11. _____

12. A triangular logo on the back of a T-shirt has a base of $5\frac{1}{2}$ inches and a height of 4 inches. What is the area of the logo?

12. _____

**Refer to the parallelogram at the right for Exercises 13 and 14. Justify your answers.**

13. Suppose the base and height are each multiplied by $\frac{1}{2}$. What effect would this have on the area?

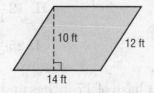

13. _____

14. Suppose the side lengths are multiplied by 2. Describe the change in the perimeter.

14. _____

15. Find the area of the figure at the right.

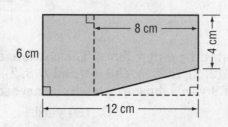

15. _____

# Test, Form 2B

**Write the letter for the correct answer in the blank at the right of each question.**

1. A package of sticky notes is in the shape of a parallelogram. The dimensions of one sticky note are shown. What is the area of one sticky note?

   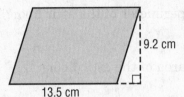

   **A.** 248.4 cm²          **C.** 62.1 cm²

   **B.** 124.2 cm²          **D.** 22.7 cm²

   1. _____

2. What is the base of a parallelogram with height 7.3 meters and an area of 65.7 square meters?

   **F.** 3.65 m          **H.** 9 m

   **G.** 4.5 m          **I.** 18 m

   2. _____

3. What is the area of the triangle?

   **A.** 578 m²

   **B.** 374 m²

   **C.** 289 m²

   **D.** 187 m²

   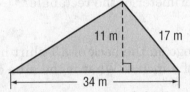

   3. _____

4. Jaida is buying a triangular-shaped rug. It has a height of $3\frac{1}{2}$ feet and a base of $4\frac{1}{4}$ feet. What is the area of the rug?

   **F.** $7\frac{7}{16}$ ft²          **H.** $12\frac{15}{16}$ ft²

   **G.** $7\frac{3}{4}$ ft²          **I.** $14\frac{7}{8}$ ft²

   4. _____

5. A triangle has a base of 15 inches and an area of 82.5 square inches. What is the height of the triangle?

   **A.** 5.5 in.          **C.** 22 in.

   **B.** 11 in.          **D.** 67.5 in.

   5. _____

6. What is the area of the trapezoid?

   **F.** 588 mm²

   **G.** 354 mm²

   **H.** 294 mm²

   **I.** 234 mm²

   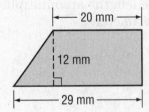

   6. _____

7. A farmer is installing a new fence. The coordinates of the vertices of the fence are $A(1, 1)$, $B(1, 9)$, $C(7, 9)$, and $D(7, 1)$. If each grid square has a length of 5 yards, how much wire is needed for the fence?

   **A.** 28 yards          **C.** 140 yards

   **B.** 48 yards          **D.** 240 yards

   7. _____

# Test, Form 2B (continued)

SCORE _____

**For Exercises 8 and 9, refer to the figure at the right
that shows the dimensions of Gabby's attic floor.**

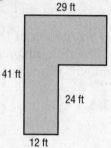

8. What is the perimeter of the attic floor?

8. _____

9. What is the area of the attic floor?

9. _____

**For Exercises 10 and 11, use the following information.
A rectangle has vertices $A(2, 6)$, $B(2, 9)$, $C(7, 9)$, and $D(7, 6)$.**

10. What is the length of each side of the rectangle?

10. _____

11. What is the perimeter of the rectangle?

11. _____

12. A triangular logo on the back of a T-shirt has a base of $7\frac{1}{2}$ inches
and a height of 4 inches. What is the area of the logo?

12. _____

**Refer to the parallelogram at the right for
Exercises 13 and 14. Justify your answers.**

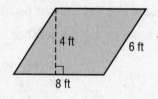

13. Suppose the base and height are each
multiplied by $\frac{1}{2}$. What effect would this have
on the area?

13. _____

14. Suppose the side lengths are multiplied by 2. Describe the
change in the perimeter.

14. _____

15. Find the area of the figure at the right.

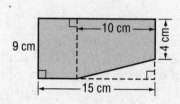

15. _____

# Test, Form 3A

**Write the correct answer in the blank at the right of each question.**

1. Find the area of the parallelogram.

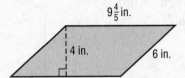

1. _____

2. Find the base of a parallelogram with height 11.8 millimeters and an area of 151.04 square millimeters.

2. _____

3. The play area in a local park has the dimensions shown at the right. What is the area of the park?

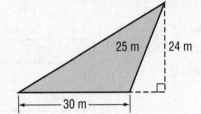

3. _____

4. A triangular logo that is painted on the side of a building has a base length of 8 feet and a height of 4 yards. What is the area of the logo in square feet?

4. _____

5. A triangle has a base of 35 centimeters and an area of 819 square centimeters. What is the height of the triangle?

5. _____

6. The figure at the right shows the approximate dimensions of Clare County. What is the area of Clare County?

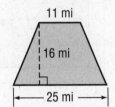

6. _____

7. Suppose the base and height of the parallelogram in Exercise 1 is multiplied by 3. What effect would this have on the area of the parallelogram? Explain your reasoning.

7. _____

8. The side lengths of a triangle are 3 meters, 4 meters, and 5 meters. Suppose the side lengths are multiplied by 4. Describe the change in the perimeter.

8. _____

# Test, Form 3A   *(continued)*

SCORE _____

9. The figure at the right shows the dimensions of the garden in Marissa's back yard. What is the area of the garden?

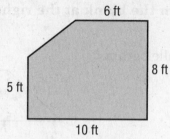

9. _____

10. A parallelogram has an area of 100 square units. Its perimeter is between 40 and 60 units. List two possible dimensions for the parallelogram.

10. _____

11. What is the area of the shaded region in the figure at the right?

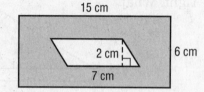

11. _____

12. Find the height of a trapezoid given that it has an area of 650 square feet and the lengths of its bases are 23 feet and 42 feet.

12. _____

13. Find the area of the figure below in square units.

13. _____

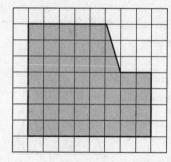

14. A figure has vertices at $A(1, 2)$, $B(2, 6)$, $C(4, 6)$, and $D(6, 2)$. Graph the figure and classify it. Then find the area.

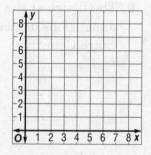

14. _____

15. A rectangle has vertices $W(2, 3)$, $X(2, 6)$, $Y(7, 6)$, and $Z(7, 3)$. Describe how to find the side lengths of the rectangle without graphing.

15. _____

# Test, Form 3B

**Write the correct answer in the blank at the right of each question.**

1. Find the area of the parallelogram.

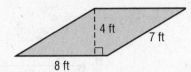

1. _____

2. Find the base of a parallelogram with height 13.5 yards and an area of 159.3 square yards.

2. _____

3. A scrap of fabric has the dimensions shown. What is the area of the scrap of fabric? Round to the nearest tenth.

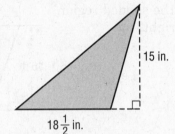

3. _____

4. A wheat cracker is shaped like a triangle with a base of 50 millimeters and a height of 5.7 centimeters. What is the area of the cracker in square centimeters? Round to the nearest tenth.

4. _____

5. A triangle has a base of 35 feet and an area of 630 square feet. What is the height of the triangle?

5. _____

6. Michael's placemat has the dimensions shown below. What is the area of the placemat?

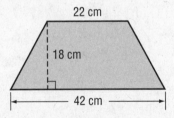

6. _____

7. Suppose the base and height of the parallelogram in Exercise 1 is multiplied by 4. What effect would this have on the area of the parallelogram? Explain your reasoning.

7. _____

8. The sides lengths of a triangle are 5 feet, 10 feet, and 13 feet. Suppose the side lengths are multiplied by 5. Describe the change in perimeter.

8. _____

# Test, Form 3B  *(continued)*

9. The figure below shows the dimensions of the local dog park. What is the area of the dog park?

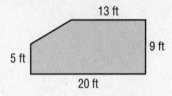

9. _____

10. A parallelogram has an area of 120 square units. Its perimeter is between 40 and 50 units. List two possible dimensions for the parallelogram.

10. _____

11. What is the area of the shaded region in the figure at the right?

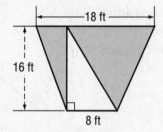

11. _____

12. Find the height of a trapezoid given that it has an area of 1,395 square feet and the lengths of its bases are 35 feet and 58 feet.

12. _____

13. Find the area of the figure below in square units.

13. _____

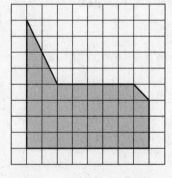

14. A figure has vertices at $A(3, 1)$, $B(3, 7)$, and $C(6, 1)$. Graph the figure and classify it. Then find the area.

14. _____

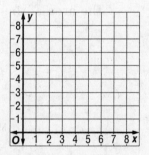

15. A rectangle has vertices $W(2, 3)$, $X(2, 6)$, $Y(7, 6)$, and $Z(7, 3)$. Describe how to find the side lengths of the rectangle without graphing.

15. _____

# Are You Ready?

## Review

---

**Order of Operations**

**1.** Simplify the expressions inside grouping symbols, like parentheses.
**2.** Find the value of all powers.
**3.** Multiply and divide in order from left to right.
**4.** Add and subtract in order from left to right.

---

## Example 1

**Evaluate $(56 \div 7) + (3 \cdot 4)$.**

$$
\begin{aligned}
(56 \div 7) + (3 \cdot 4) &= 8 + (3 \cdot 4) \qquad &\text{Divide 56 by 7.} \\
&= 8 + 12 &\text{Multiply 3 by 4.} \\
&= 20 &\text{Add.}
\end{aligned}
$$

## Exercises

**Evaluate.**

**1.** $(64 \div 8) + (9 + 7)$      1. _____

**2.** $37 + 42 - 28$      2. _____

**3.** $(13 + 7) \div (40 \div 8)$      3. _____

**4.** $32 + 16 - 40$      4. _____

**5.** $(16 + 3) + (5 - 2)$      5. _____

**6.** $(6 + 8) + (8 \cdot 2)$      6. _____

**7.** $(42 \div 7) + (4 - 1)$      7. _____

**8.** $9 + (54 \div 9)$      8. _____

**9.** $(49 + 4) - (11 + 2)$      9. _____

**10.** $(4 \times 6) + (3 \times 5)$      10. _____

**11.** $(20 - 2) \div (3 \times 2)$      11. _____

**12.** $(45 \div 9) \times (8 + 3)$      12. _____

# Are You Ready?

## Practice

**Evaluate.**

1. $(49 - 1) \div (10 + 6)$

2. $(38 + 51) - (29 + 2)$

3. $15 + 14 - 19$

4. $(30 + 3) + (6 - 2)$

5. $(37 + 17) \div (3 \times 3)$

6. $(24 \div 6) + (5 - 3)$

7. $(8 + 18) \div (9 + 4)$

8. $11 + (36 \div 4)$

**Multiply.**

9. $3 \times 4 \times 2.1$

10. $2 \times 2.7 \times 3$

11. $4 \times 1.1 \times 2$

12. $1.7 \times 3 \times 2$

13. $8.1 \times 4 \times 2$

14. $7.5 \times 3 \times 1$

15. $10.1 \times 4 \times 5$

16. $8 \times 3.7 \times 4$

17. $7 \times 2.9 \times 1$

18. $8 \times 3.4 \times 9$

19. $7 \times 5.5 \times 2$

20. $6.4 \times 3 \times 2$

1. _____

2. _____

3. _____

4. _____

5. _____

6. _____

7. _____

8. _____

9. _____

10. _____

11. _____

12. _____

13. _____

14. _____

15. _____

16. _____

17. _____

18. _____

19. _____

20. _____

# Are You Ready?

## Apply

| | |
|---|---|
| **1. BASEBALL** John had 30 baseball cards. He gave 14 cards to Mike and 7 to Jeff. Then he bought 6 more cards. How many baseball cards does John have now? | **2. PARTIES** Louise is making a chicken dish for 6 people. The recipe for the chicken dish she is making calls for 1.2 pounds of chicken for each person. Louise wants to double the recipe. How many pounds of chicken will she need? |
| **3. FUNDRAISER** The soccer team collected soda cans for a fundraiser. They had 175 cans and found 58 more. The next day, they turned in 97 cans. How many cans do they have left? | **4. CORN** Mr. Rodriguez planted 22 rows of corn. There were 15 plants in each row. He also planted 5 rows of tomato plants with each row havin 12 plants. How many plants did he plant in all? |
| **5. EARNINGS** Max earns $7.25 an hour. If he works for 6 hours a week for 3 weeks, how much will he earn? | **6. WALKING** Val can walk 3.2 miles in an hour. If she walks for 4 hours a week for 5 weeks, how many miles will she walk? |

# Diagnostic Test

## Evaluate.

1. $(37 + 52) - (50 - 4)$      1. _____

2. $(5 + 11 \times 4) \div 7$      2. _____

3. $36 \div (3 + 1)$      3. _____

4. $(7 + 8) + (54 \div 9)$      4. _____

5. $(6 + 40) \div (8 \div 4)$      5. _____

6. $(30 \div 5) \div (4 - 2)$      6. _____

7. $50 + 10 - 41$      7. _____

8. $(12 + 7) + (5 - 2)$      8. _____

## Multiply.

9. $12 \times 4.1 \times 2$      9. _____

10. $13 \times 3.2 \times 3$      10. _____

11. $11 \times 2.1 \times 10$      11. _____

12. $9 \times 8.1 \times 7$      12. _____

13. $3.2 \times 4 \times 3$      13. _____

14. $5.7 \times 6 \times 7$      14. _____

15. $5.5 \times 9 \times 8$      15. _____

16. **WORK** Anna earns \$10.75 an hour. If she works 12 hours a week for 6 weeks, how much will she make?      16. _____

17. **BASKETBALL** The basketball team is selling raffle tickets to buy new uniforms. The team earns \$5.50 for each ticket they sell. How much will the team earn if each player sells 15 tickets and there are 8 players on the team?      17. _____

# Pretest

**Find the volume of each figure.**

**1.**

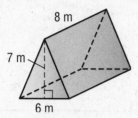

8 m

7 m

6 m

1. _____

**2.**

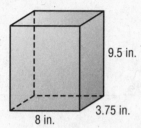

9.5 in.

8 in.

3.75 in.

2. _____

**3.**

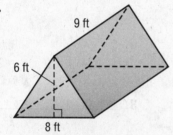

9 ft

6 ft

8 ft

3. _____

**4.**

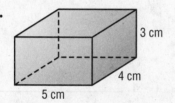

3 cm

4 cm

5 cm

4. _____

**Find the surface area of the figure.**

**5.**

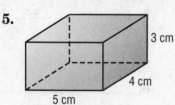

3 cm

4 cm

5 cm

5. _____

# Chapter Quiz

**Find the volume of each prism. Round to the nearest tenth if necessary.**

**1.**

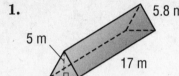

5.8 m

5 m

17 m

6 m

1. _____

**2.**

5.4 in.

6 in.

7.2 in.

2. _____

**3.**

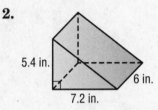

2 ft

16.2 ft

16 ft

3. _____

**4.**

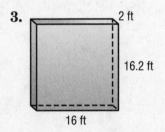

25 mr

4 mm

22 mm

4. _____

**5. RECREATION** The base of a skateboard ramp has an area of 2.3 square meters. The height of the ramp is 1.2 meters. Find the volume of the triangular prism.

5. _____

**6.** Find the missing dimension of a rectangular prism with a volume of 120 cubic feet, a width of 4 feet, and a length of 4 feet.

6. _____

**7.** A rectangular prism has a volume of 2,288 cubic meters, a height of 13 meters, and a length of 22 meters. What is the measure of the missing dimension?

7. _____

# Vocabulary Test

SCORE _____

| | | |
|---|---|---|
| cubic units | slant height | triangular prism |
| lateral face | surface area | vertex |
| rectangular prism | three-dimensional figure | volume |

## Write *true* or *false* for each statement.

1. Surface area is the sum of the areas of all the surfaces of a three-dimensional object.

1. _____

2. A triangular prism has exactly one base.

2. _____

3. The volume of a triangular prism can be found using the formula $V = Bh$.

3. _____

4. The volume of a rectangular prism is half the volume of a triangular prism with the same base area and height.

4. _____

5. A triangular prism always has three faces that are triangles.

5. _____

6. The volume of a triangular prism is one third the volume of a rectangular prism with the same base area and height.

6. _____

7. A net of a triangular prism shows that it has three rectangular faces.

7. _____

## Define in your own words.

8. volume

8. _____

9. lateral face

9. _____

# Standardized Test Practice

Read each question. Then fill in the correct answer on the answer sheet provided by your teacher or on a sheet of paper.

1. Olivia is placing a gift inside a box that measures 15 centimeters by 8 centimeters by 3 centimeters. What is the surface area of the box?

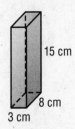

15 cm
8 cm
3 cm

A. 378 square centimeters
B. 288 square centimeters
C. 189 square centimeters
D. 26 square centimeters

2. Sandi drinks the same amount of water every week as shown in the table.

| Gallons of Water | 8.5 | 17 | 25.5 | 34 | ▪ |
|---|---|---|---|---|---|
| Number of Weeks | 1 | 2 | 3 | 4 | 5 |

How many gallons of water does she drink in 5 weeks?

F. 13.5
G. 42
H. 42.5
I. 48

3. Rigo is buying a subscription to a sports magazine that costs $2.19 per issue. Which of the following is a good estimate for the cost of 12 issues?

A. $12
B. $18
C. $24
D. $36

4. ✏️ **GRIDDED RESPONSE** Mail Magic sells two different rectangular mailing boxes with the dimensions shown in the table. What is the volume of Box B in cubic inches?

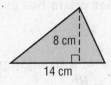

|  | Length (in.) | Width (in.) | Height (in.) |
|---|---|---|---|
| Box A | 16 | 10 | 8 |
| Box B | 10 | 8 | 8 |

5. What is the area of the triangle?

8 cm
14 cm

F. 22 cm²
G. 44 cm²
H. 56 cm²
I. 112 cm²

6. ▪ **SHORT RESPONSE** What is the area of the parallelogram shown below?

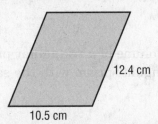

12.4 cm
10.5 cm

7. Which function rule describes the relationship between $x$, time traveled, and $y$, distance from home?

| Time (h) | $x$ | 0 | 1 | 2 | 3 |
|---|---|---|---|---|---|
| Distance from Home (mi) | $y$ | 0 | 65 | 130 | 195 |

A. $65y = x$
B. $y = 65 \div x$
C. $y = x + 65$
D. $y = 65x$

NAME _____ DATE _____ PERIOD _____

**8.** ▦✎ **GRIDDED RESPONSE** What is the area of trapezoid shown below in square feet?

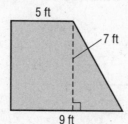

5 ft

7 ft

9 ft

**9.** Super Toys makes two sizes of cube-shaped building blocks. The larger block has side lengths four times the side lengths of the smaller block. What is the ratio of the surface area of the small block to the surface area of the large block?

**F.** 1 to 4
**G.** 1 to 6
**H.** 1 to 16
**I.** 1 to 32

**10.** ▦ **SHORT RESPONSE** Find the area of the trapezoid.

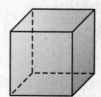

3 in.

5 in.

11 in.

**11.** A rectangular prism has a surface area of 254 square meters, a height of 9 meters, and a length of 7 meters. What is the measure of the width of the prism?

**A.** 4 m     **C.** 6 m
**B.** 5 m     **D.** 7 m

**12.** A rectangular prism has a volume of 60 cubic feet, a width of 2 feet, and a height of 3 feet. What is the measure of the length of the prism?

**F.** 14 ft
**G.** 13 ft
**H.** 12 ft
**I.** 10 ft

**13.** The table shows the cost of renting a booth at a fall festival. There is an initial charge for reserving a booth and a fee per day. What is the cost in dollars of renting a booth for all 7 days of the festival?

| Days | 0 | 1 | 2 | 3 |
|---|---|---|---|---|
| Cost ($) | 50 | 90 | 130 | 170 |

**A.** $350
**B.** $330
**C.** $280
**D.** $70

**14.** ▦✎ **GRIDDED RESPONSE** A rectangle has an area of 318 square centimeters and a width of 12 centimeters. What is the length of the rectangle in centimeters?

**15.** ▦ **EXTENDED RESPONSE** Leora is gift-wrapping the box shown.

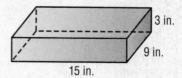

3 in.
9 in.
15 in.

**Part A** Find the volume of the box.

**Part B** If each dimension is doubled, explain what happens to the volume.

**Part C** If only one dimension is doubled, what happens to the volume? Does it matter which dimension is doubled? Explain.

# Student Recording Sheet

*Use this recording sheet with the Standardized Test Practice pages.*

**Fill in the correct answer. For gridded-response questions, write your answers in the boxes on the answer grid and fill in the bubbles to match your answers.**

1. Ⓐ Ⓑ Ⓒ Ⓓ

2. Ⓕ Ⓖ Ⓗ Ⓘ

3. Ⓐ Ⓑ Ⓒ Ⓓ

4.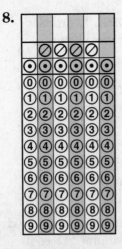

5. Ⓕ Ⓖ Ⓗ Ⓘ

6. _____

7. Ⓐ Ⓑ Ⓒ Ⓓ

8.

9. Ⓕ Ⓖ Ⓗ Ⓘ

10. _____

11. Ⓐ Ⓑ Ⓒ Ⓓ

12. Ⓕ Ⓖ Ⓗ Ⓘ

13. Ⓐ Ⓑ Ⓒ Ⓓ

14.

## Extended Response

Record your answers for Exercise 15 on the back of this paper.

**Course 1 • Chapter 10** Volume and Surface Area

# Extended-Response Test

Demonstrate your knowledge by giving a clear, concise solution to each problem. Be sure to include all relevant drawings and justify your answers. You may show your solution in more than one way or investigate beyond the requirements of the problem. If necessary, record your answer on another piece of paper.

1. The manager of a cereal company is going to package a new cereal in a rectangular prism that measures 2 inches by 6 inches by 8 inches.

    **a.** How much material will he need to produce it?

    **b.** How much would the container hold?

    **c.** How would your answer to part **b** change if the height of the container was divided by 2?

2. A toy building kit contains plastic solids in a variety of shapes. One of these shapes is shown at right. The piece has a volume of 13 cubic centimeters. Find the height of the triangular bases.

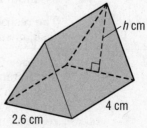

3. A baby toy is in the shape of pyramid. The pyramid has all sides that are equilateral triangles. Each triangle has side lengths of 12 inches. The slant height is 10.4 inches.

    **a.** Draw a net of the pyramid.

    **b.** Explain how you would find the surface area of the toy.

    **c.** Find the surface area of the toy. Show how you solved.

# Extended-Response Rubric

| Score | Description |
|-------|-------------|
| 4 | A score of four is a response in which the student demonstrates a thorough understanding of the mathematics concepts and/or procedures embodied in the task. The student has responded correctly to the task, used mathematically sound procedures, and provided clear and complete explanations and interpretations.<br><br>The response may contain minor flaws that do not detract from the demonstration of a thorough understanding. |
| 3 | A score of three is a response in which the student demonstrates an understanding of the mathematics concepts and/or procedures embodied in the task. The student's response to the task is essentially correct with the mathematical procedures used and the explanations and interpretations provided demonstrating an essential but less than thorough understanding.<br><br>The response may contain minor flaws that reflect inattentive execution of mathematical procedures or indications of some misunderstanding of the underlying mathematics concepts and/or procedures. |
| 2 | A score of two indicates that the student has demonstrated only a partial understanding of the mathematics concepts and/or procedures embodied in the task. Although the student may have used the correct approach to obtaining a solution or may have provided a correct solution, the student's work lacks an essential understanding of the underlying mathematical concepts.<br><br>The response contains errors related to misunderstanding important aspects of the task, misuse of mathematical procedures, or faulty interpretations of results. |
| 1 | A score of one indicates that the student has demonstrated a very limited understanding of the mathematics concepts and/or procedures embodied in the task. The student's response is incomplete and exhibits many flaws. Although the student's response has addressed some of the conditions of the task, the student reached an inadequate conclusion and/or provided reasoning that was faulty or incomplete.<br><br>The response exhibits many flaws or may be incomplete. |
| 0 | A score of zero indicates that the student has provided no response at all, or a completely incorrect or uninterpretable response, or demonstrated insufficient understanding of the mathematics concepts and/or procedures embodied in the task. For example, a student may provide some work that is mathematically correct, but the work does not demonstrate even a rudimentary understanding of the primary focus of the task. |

NAME _____ DATE _____ PERIOD _____

# Test, Form 1A

SCORE _____

*Write the letter for the correct answer in the blank at the right of each question.*

1. A rectangular inflatable swimming pool is $2\frac{4}{5}$ yards long, 2 yards wide, and $\frac{3}{5}$ yard tall. What is the volume of the pool? Round to the nearest tenth.
   A. 3.4 yd³
   B. 5.4 yd³
   C. 17.0 yd³
   D. 22.0 yd³

1. _____

2. What is the surface area of the rectangular prism below?

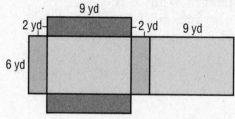

   F. 17 yd²
   G. 84 yd²
   H. 108 yd²
   I. 168 yd²

2. _____

**For Exercises 3 and 4, find the volume of each figure. Round to the nearest tenth if necessary.**

3. A. 28.9 in³
   B. 43.3 in³
   C. 57.8 in³
   D. 86.7 in³

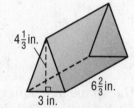

3. _____

4. F. 824.8 ft³
   G. 530.8 ft³
   H. 412.4 ft³
   I. 28.3 ft³

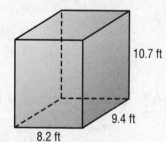

4. _____

5. A storage container is in the shape of a rectangular prism. The container has a length of 5 feet, a width of 9 feet, and a height of 8 feet. What is the surface area of the container?
   A. 360 ft³     B. 314 ft²     C. 157 ft²     D. 22 ft²

5. _____

6. What is the surface area of the triangular prism?
   F. 150 in²
   G. 180 in²
   H. 240 in²
   I. 3,120 in²

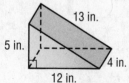

6. _____

# Test, Form 1A    (continued)

**7.** What is the volume of the prism?

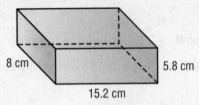

8 cm   5.8 cm   15.2 cm

**A.** 29 cm³

**B.** 87 cm³

**C.** 512.3 cm³

**D.** 705.3 cm³

**7.** _____

**8.** A department store wraps gifts for its customers. What is the amount of wrapping paper it would take to wrap a gift box with a width of 12 inches, a height of 4 inches, and a length of 16 inches?

**F.** 768 in²

**G.** 608 in²

**H.** 512 in²

**I.** 304 in²

**8.** _____

**9.** A triangular prism has the dimensions shown at the right. What is the volume of the prism? Round to the nearest tenth if necessary.

**A.** 187.9 mm³

**B.** 94.0 mm³

**C.** 47.0 mm³

**D.** 42.7 mm³

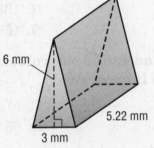

6 mm   5.22 mm   3 mm

**9.** _____

**10.** A paperweight in the shape of a square pyramid has a side length of 2.7 inches. The slant height is 2.5 inches. What is the surface area of the paper weight?

**F.** 18.2 in²

**G.** 19.75 in²

**H.** 20.79 in²

**I.** 34.29 in²

**10.** _____

**11.** What is the surface area of the triangular prism?

**A.** 216 m²

**B.** 240 m²

**C.** 264 m²

**D.** 312 m²

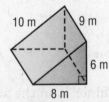

10 m   9 m   6 m   8 m

**11.** _____

**12.** What is the surface area of the square pyramid?

**F.** 160 in²

**G.** 240 in²

**H.** 340 in²

**I.** 1,200 in²

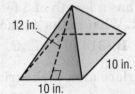

12 in.   10 in.   10 in.

**12.** _____

# Test, Form 1B

*Write the letter for the correct answer in the blank at the right of each question.*

1. A rectangular inflatable swimming pool is 3 yards long, $2\frac{4}{5}$ yards wide, and $1\frac{2}{5}$ yards tall. What is the volume of the pool? Round to the nearest tenth.
   - **A.** 7.2 yd³
   - **B.** 11.8 yd³
   - **C.** 27 yd³
   - **D.** 33.0 yd³

   1. _____

2. What is the surface area of the rectangular prism below?

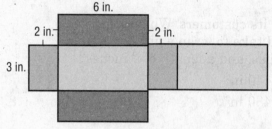

   - **F.** 72 in²
   - **G.** 144 in²
   - **H.** 720 in²
   - **I.** 1,440 in²

   2. _____

**For Exercises 3 and 4, find the volume of each figure. Round to the nearest tenth if necessary.**

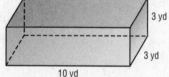

3. **A.** 90 yd³
   **B.** 60 yd³
   **C.** 45 yd³
   **D.** 30 yd³

   3. _____

4. **F.** 40.8 in³
   **G.** 20.4 in³
   **H.** 13.6 in³
   **I.** 2.4 in³

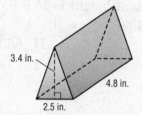

   4. _____

5. A storage container is in the shape of a rectangular prism. The container has a length of 10 feet, a width of 8 feet, and a height of 12 feet. What is the surface area of the container?
   - **A.** 960 ft²
   - **B.** 592 ft²
   - **C.** 96 ft²
   - **D.** 60 ft²

   5. _____

6. What is the surface area of the triangular prism?
   - **F.** 168 in²
   - **G.** 216 in²
   - **H.** 264 in²
   - **I.** 3,360 in²

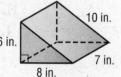

   6. _____

# Test, Form 1B   (continued)

7. What is the volume of the rectangular prism? Round to the nearest tenth if necessary.

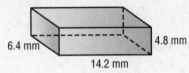

A. 25.4 mm³                    C. 379.5 mm³

B. 76.2 mm³                    D. 436.2 mm³

7. _____

8. A department store wraps gifts for its customers. What is the amount of wrapping paper it would take to wrap a gift box with a width of 8 inches, a height of 6 inches, and a length of 5 inches?

F. 236 in²                     H. 440 in²

G. 240 in²                     I. 480 in²

8. _____

9. A triangular prism has the dimensions shown at the right. What is the volume of the prism? Round to the nearest tenth if necessary.

A. 156 cm³                     C. 315 cm³

B. 195 cm³                     D. 630 cm³

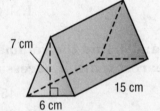

9. _____

10. A paperweight in the shape of a square pyramid has a side length of 8.2 inches. The slant height is 15.9 inches. What is the surface area of the paper weight?

F. 328.0 in²                   H. 588.8 in²

G. 513.6 in²                   I. 1,069.1 in²

10. _____

11. What is the surface area of the triangular prism?

A. 612 m²                      C. 756 m²

B. 720 m²                      D. 1,512 m²

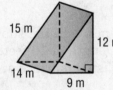

11. _____

12. What is the surface area of the square pyramid?

F. 294 cm²                     H. 133 cm²

G. 217 cm²                     I. 70 cm²

12. _____

# Test, Form 2A

**For Exercises 1 and 2, find the volume of each figure. Round to the nearest tenth if necessary.**

**1.**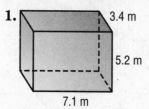

3.4 m

5.2 m

7.1 m

**A.** 251.1 m³

**B.** 214 m³

**C.** 125.5 m³

**D.** 25 m³

1. _____

**2.**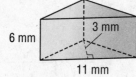

6 mm

3 mm

11 mm

**F.** 33 mm³

**G.** 99 mm³

**H.** 198 mm³

**I.** 231 mm³

2. _____

**3.** The lid of a jewelry box is in the shape of a triangular prism. The lid has a height of 10 inches. The triangular base of the lid has a base of $6\frac{1}{2}$ inches and a height of $3\frac{1}{2}$ inches. What is the volume of the lid to the nearest tenth?

**A.** 277.5 in³

**B.** 113.8 in³

**C.** 98.8 in³

**D.** 20 in³

3. _____

**For Exercises 4 and 5, find the surface area of each solid. Round to the nearest tenth if necessary.**

**4.**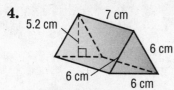

5.2 cm

7 cm

6 cm

6 cm

6 cm

**F.** 109.2 cm²

**G.** 157.2 cm²

**H.** 188.4 cm²

**I.** 218.4 cm²

4. _____

**5.**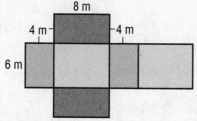

8 m

4 m

4 m

6 m

**A.** 768 m²

**B.** 208 m²

**C.** 192 m²

**D.** 104 m²

5. _____

# Test, Form 2A   *(continued)*

6. Ming has four rectangular baking pans. Which of the pans has the greatest volume?

   **F.** Pan A: 9 in. by 16 in. by $1\frac{1}{2}$ in.

   **G.** Pan B: 9 in. by 13 in. by 1 in.

   **H.** Pan C: 9 in. by 9 in. by 2 in.

   **I.** Pan D: 8 in. by 8 in. by $2\frac{1}{2}$ in.

6. _____

7. A room is 30 feet long, 25 feet wide, and 10 feet tall. If Mickey paints the walls and the ceiling, how much surface area will he cover?

   **A.** 2,600 ft²

   **B.** 2,350 ft²

   **C.** 2,300 ft²

   **D.** 1,850 ft²

7. _____

8. A rectangular prism has a volume of 468 cubic meters. It has a length of 12 meters and a width of 6 meters. What is the height of the prism?

8. _____

**For Exercises 9 and 10, find the surface area of each figure. Round to the nearest tenth if necessary.**

9.
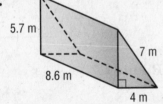

9. _____

10.
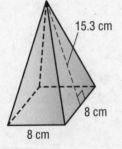

10. _____

11. A rectangular prism has a length of $1\frac{1}{2}$ feet, a width of $2\frac{1}{2}$ feet, and a height of 8 inches. What is the volume of the prism in cubic inches?

11. _____

12. A pyramid has all sides that are equilateral triangles. Each triangle has side lengths of 9 centimeters. If the surface area of the pyramid is 140.4 square centimeters, what is the slant height of the pyramid?

12. _____

# Test, Form 2B

**For Exercises 1 and 2, find the volume of each figure.**
**Round to the nearest tenth if necessary.**

**1.**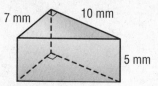

7 mm   10 mm

5 mm

    **A.** 35 mm³             **C.** 175 mm³

    **B.** 125 mm³          **D.** 350 mm³

1. _____

**2.**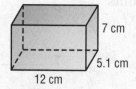

7 cm

5.1 cm

12 cm

    **F.** 428.4 cm³         **H.** 42.8 cm³

    **G.** 214.2 cm³        **I.** 21.4 cm³

2. _____

**3.** The lid of a jewelry box is in the shape of a triangular prism. The lid has a height of 12 inches. The triangular base of the lid has a base of $8\frac{1}{2}$ inches and a height of $4\frac{1}{2}$ inches. What is the volume of the lid to the nearest tenth?

    **A.** 459.0 in³         **C.** 75.0 in³

    **B.** 229.5 in³        **D.** 25.0 in³

3. _____

**For Exercises 4 and 5, find the surface area of each solid.**
**Round to the nearest tenth if necessary.**

**4.**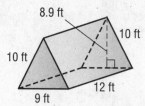

8.9 ft

10 ft

10 ft

9 ft    12 ft

    **F.** 49.9 ft²     **G.** 388.1 ft²     **H.** 428.1 ft²     **I.** 480.6 ft²

4. _____

**5.**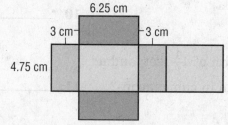

6.25 cm

3 cm    3 cm

4.75 cm

    **A.** 12.5 cm²     **B.** 89.1 cm²     **C.** 125.4 cm²     **D.** 250.8 cm²

5. _____

# Test, Form 2B    *(continued)*

6. Pat has four rectangular baking pans. Which of the pans has the greatest volume?

   **F.** Pan A: 8 in. by 15 in. by $1\frac{1}{2}$ in.

   **G.** Pan B: 9 in. by 13 in. by 2 in.

   **H.** Pan C: 8 in. by 8 in. by $1\frac{1}{2}$ in.

   **I.** Pan D: 11 in. by 14 in. by $\frac{1}{2}$ in.

6. _____

7. A room is 20 feet long, 18 feet wide, and 8 feet tall. If Keiko paints the walls and the ceiling, how much surface area will she cover?

   **A.** 968 ft²            **C.** 1,184 ft²

   **B.** 1,168 ft²         **D.** 1,328 ft²

7. _____

8. A rectangular prism has a volume of 810 cubic meters. It has a length of 10 meters and a width of 9 meters. What is the height of the prism?

8. _____

**For Exercises 9 and 10, find the surface area of each figure. Round to the nearest tenth if necessary.**

9.

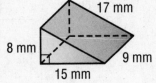

9. _____

10.

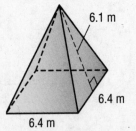

10. _____

11. A rectangular prism has a length of $2\frac{3}{4}$ feet, a width of $4\frac{1}{2}$ feet, and a height of 6 inches. What is the volume of the prism in cubic inches?

11. _____

12. A pyramid has all sides that are equilateral triangles. Each triangle has side lengths of 11 centimeters. If the surface area of the pyramid is 330 square centimeters, what is the slant height of the pyramid?

12. _____

# Test, Form 3A

**For Exercises 1 and 2, find the volume of each figure. Round to the nearest tenth if necessary.**

**1.**

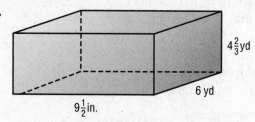

$4\frac{2}{3}$ yd

6 yd

$9\frac{1}{2}$ in.

1. _____

**2.**

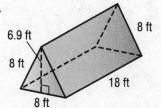

6.9 ft

8 ft

8 ft

8 ft

18 ft

2. _____

**For Exercises 3 and 4, find the surface area of each solid.**

**3.**

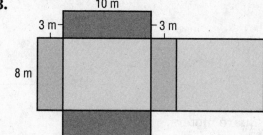

10 m

3 m        3 m

8 m

3. _____

**4.**

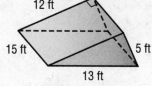

12 ft

15 ft        5 ft

13 ft

4. _____

**5.**

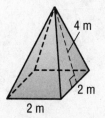

4 m

2 m

2 m

5. _____

**6.** A gift box in the shape of a triangular prism has a volume of 140 cubic inches, a base height of 10 inches, and a height of 4 inches. What is the length of the base?

6. _____

# Test, Form 3A  (continued)

7. A special box designed to hold an antique artifact is shaped like a triangular prism. The surface area of the box is 421.2 square inches. The height of the base triangle is 7.8 inches and each side of the base triangle is 9 inches long. What is the height of box?

7. _____

8. The dimensions of two different-sized boxes of popcorn are shown in the table. Which box holds more popcorn? How much more?

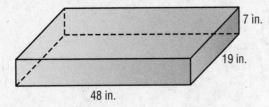

| Boxes of Popcorn | | | |
|---|---|---|---|
| Box | Length (in.) | Width (in.) | Height (in.) |
| A | $8\frac{1}{2}$ | 4 | $12\frac{3}{4}$ |
| B | $10\frac{1}{2}$ | $6\frac{3}{4}$ | 6 |

8. _____

9. Find the volume and surface area of the prism.

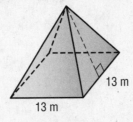

7 in.

19 in.

48 in.

9. _____

10. The surface area of the pyramid below is 533 square meters. What is the slant height of the pyramid?

13 m

13 m

10. _____

11. Piper wants to buy enough potting soil to fill a flower box that is 36 inches long, 8 inches wide, and 10 inches tall. If one bag of potting soil contains 576 cubic inches, how many bags should she buy?

11. _____

12. Michael's pencil box is shaped like a rectangular prism. It is 30.5 centimeters long, 20.3 centimeters wide, and 10 centimeters tall. He estimates that the surface area of the box is about 2,200 square centimeters. Is his estimate reasonable? Explain your reasoning.

12. _____

# Test, Form 3B

**For Exercises 1 and 2, find the volume of each figure. Round to the nearest tenth if necessary.**

**1.**

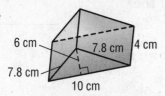

6 cm  7.8 cm  4 cm
7.8 cm
10 cm

**1.** _____

**2.**

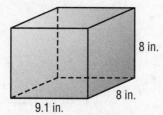

8 in.
8 in.
9.1 in.

**2.** _____

**For Exercises 3–5, find the surface area of each figure.**

**3.**

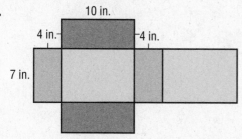

10 in.
4 in.   4 in.
7 in.

**3.** _____

**4.**

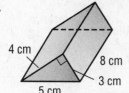

4 cm
8 cm
3 cm
5 cm

**4.** _____

**5.**

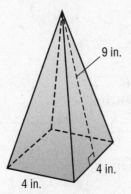

9 in.
4 in.
4 in.

**5.** _____

**6.** A gift box in the shape of a triangular prism has a volume of 36 cubic inches, a base height of 4 inches, and a height of 3 inches. What is the length of the base?

**6.** _____

# Test, Form 3B  (continued)

7. A special box designed to hold an antique artifact is shaped like a triangular prism. The surface area of the box is 439.2 square inches. The height of the base triangle is 6.9 inches and each side of the base triangle is 8 inches long. What is the height of box?

7. _____

8. The dimensions of two different-sized boxes of cereal are shown in the table. Which box holds more cereal? How much more?

| Boxes of Cereal | | | |
|---|---|---|---|
| Box | Length (in.) | Width (in.) | Height (in.) |
| A | $10\frac{1}{4}$ | 3 | $14\frac{1}{2}$ |
| B | $11\frac{1}{2}$ | 5 | $12\frac{3}{4}$ |

8. _____

9. Find the volume and surface area of the prism.

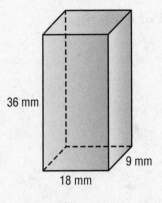

36 mm

9 mm

18 mm

9. _____

10. The surface area of the pyramid below is 175 square meters. What is the slant height of the pyramid?

7 m          7 m

10. _____

11. Piper wants to buy enough potting soil to fill a flower box that is 30 inches long, 12 inches wide, and 8 inches tall. If one bag of potting soil contains 576 cubic inches, how many bags should she buy?

11. _____

12. A shoe box is shaped like a rectangular prism. It is 39.4 centimeters long, 17.8 centimeters wide, and 11.4 centimeters tall. Jacob estimates that the surface area of a shoe box is about 2,800 square centimeters. Is his estimate reasonable? Explain your reasoning.

12. _____

# Are You Ready?

## Review

### Example

**Find 10.51 + 4.12 + 11.24 + 3.12.**

```
  10.51        Add.
   4.12
  11.24
+  3.12
  28.99
```

**Find each sum.**

**1.** 5.32 + 21.59 + 12.79 + 4.85

**2.** 4.57 + 1.59 + 31.51 + 8.12

**3.** 14.20 + 22.34 + 19.99 + 3.50

**4.** 20.49 + 18.10 + 5.75 + 37.90

**5.** 31.50 + 1.59 + 14.77 + 25.65

**1.** _____

**2.** _____

**3.** _____

**4.** _____

**5.** _____

**6. SCHOOL DANCE** Tonya went to the school dance. She paid $3.50 for admission. She bought a drink for $0.75, a bag of pretzels for $0.50, and a slice of pizza for $1.35. How much did she spend in all?

**6.** _____

**7. RAINFALL** The table shows the amount of rainfall a local city received this spring.  How much rain did the city receive for months March through June?

**7.** _____

| Rainfall | |
|---|---|
| **Month** | **Number of Inches** |
| March | 12.6 |
| April | 23.5 |
| May | 17.9 |
| June | 8.4 |

**8. SHOPPING** Sammy bought a glue stick for $2.95, a pair of scissors for $5.79, a scrapbook for $19.99, and a package of scrapbook paper for $12.50. How much did she spend in all, not including tax?

**8.** _____

# Are You Ready?

## Practice

**TRACK** For Exercises 1-3, refer to the table. The table shows the number of miles the track team ran this week.

| Running Log | |
|---|---|
| **Day** | **Number of Miles** |
| Monday | 24.3 |
| Tuesday | 27.25 |
| Wednesday | 35.9 |
| Thursday | 19.40 |
| Friday | 18.7 |

1. How many miles did the team run Monday through Wednesday?

   1. _____

2. How many miles did the team run Wednesday through Friday?

   2. _____

3. Find the total number of miles the team ran for the entire week.

   3. _____

**Find each quotient.**

4. $62.4 \div 4$

   4. _____

5. $87.5 \div 7$

   5. _____

6. $42.4 \div 4$

   6. _____

7. $30.3 \div 3$

   7. _____

8. $72.6 \div 2$

   8. _____

9. $758.4 \div 6$

   9. _____

10. The Augello family drove 564.8 miles on their vacation. They drove the same amount each of the 4 days. How many miles did they drive each day?

    10. _____

**Course 1 • Chapter 11** Statistical Measures

# Are You Ready?

## Apply

SCHOOL STORE For Exercises 1-3, refer to the table. The table shows the cost of items at the school store.

| School Store | |
|---|---|
| **Item** | **Cost ($)** |
| T-Shirt | 12.95 |
| Sweatshirt | 24.50 |
| Pennant | 8.50 |
| Flag | 14.99 |
| Button | 2.65 |

**1.** Jessica bought two sweatshirts and two flags. What was her total cost?

**2.** A new student bought a T-shirt, pennant, flag, and button. How much did the new student spend in all?

**3.** Suppose you bought one of each of the items in listed in table. What would be the total cost?

**4.** Three friends decide to split the cost of their bill evenly. If the bill was $65.79, how much will each friend pay?

**5.** The total weight of five packages of candy is 182.5 ounces. If each package of candy weighs the same amount, how much does one package weigh?

**6.** The total cost of six tickets to a music concert is $208.50. Each ticket costs the same amount. What is the cost of one ticket?

# Diagnostic Test

**Find each sum.**

1. 14.58 + 17.59 + 12.79 + 34.85

1. _____

2. 15.66 + 11.93 + 34.10 + 2.12

2. _____

3. 5.70 + 44.34 + 29.85 + 4.50

3. _____

4. 11.99 + 38.10 + 17.75 + 21.23

4. _____

5. 24.50 + 6.79 + 15.49 + 9.41

5. _____

6. **HOCKEY** Andrew went to the hockey game. He paid
$13.50 for admission. He bought a program for $4.75,
a bag of popcorn for $3.50, and a large drink for $3.35.
How much did he spend in all?

6. _____

7. **SNOWFALL** The table shows the amount of snowfall
a local city received this winter.

| Snowfall | |
|---|---|
| **Month** | **Number of Inches** |
| December | 18.6 |
| January | 27.9 |
| February | 32.4 |
| March | 12.5 |

How much snow did the city receive for months
December through March?

7. _____

8. **TICKETS** The total cost of eight tickets to a water park is
$158.80. Each ticket costs the same amount. What is the
cost of one ticket?

8. _____

9. **MONEY** Four siblings decide to split the cost of their
mother's gift evenly. If the mother's gift costs $75.40,
how much will each sibling pay?

9. _____

10. **VACATION** The Cooper family drove 987.6 miles on their
vacation. They drove the same number of miles each day.
If the trip was 6 days, how many miles did they drive
each day?

10. _____

# Pretest

**Find the mean, median, and mode for each data set.**

1. Number of pages read: 11, 8, 2, 10, 11

   1. _____

2. Points scored per volleyball game: 2, 0, 4, 8, 3, 4

   2. _____

3. ages of children taking ceramics classes:
   8, 10, 9, 12, 12, 15, 11

   3. _____

4. points scored in basketball games:
   14, 37, 28, 56, 47, 37, 43, 50

   4. _____

**The table shows the top speeds of sixteen roller coasters.**

| Top Speeds of Roller Coasters (mph) | | | |
|---|---|---|---|
| 80 | 40 | 48 | 70 |
| 93 | 95 | 60 | 40 |
| 57 | 40 | 65 | 90 |
| 92 | 72 | 60 | 72 |

5. Find the range.

   5. _____

6. Find the median, first quartile, third quartile, and interquartile range.

   6. _____

7. The table shows the weights of various dogs that visited a vet's office one day.

| Dog's Weight (lb) | |
|---|---|
| 14 | 54 |
| 27 | 45 |
| 19 | 27 |

   What is the mean absolute deviation of the data? Round to the nearest whole number.

   7. _____

8. The number of text messages Kumar sent this week are shown below.

   70, 50, 54, 89, 67

   What is the mean absolute deviation of the data?

   8. _____

# Chapter Quiz

**For Exercises 1 and 2, use the dot plot which shows the prices for different pairs of sunglasses.**

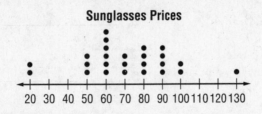

Sunglasses Prices

20 30 40 50 60 70 80 90 100 110 120 130

1. What is the range of the data?                1. _____

2. Find the mean, median, and mode.              2. _____

## Find the mean for each set of data.

3. cost of tennis shoes: $57, $63, $60, $59, $61    3. _____

4. ages of students' dogs: 5, 6, 3, 4, 8, 10, 6     4. _____

5. weights of cats at a pet store in pounds: 5, 4, 10, 7, 6    5. _____

6. cost of DVDs: $10, $5, $19, $25, $15            6. _____

## For Exercises 7–9, use the following set of data.

**TEMPERATURE** The temperature in Georgia was recorded for 6 days in November. The temperatures were 62°, 60°, 70°, 78°, 60°, and 66°.

7. What is the mean for the data?                 7. _____

8. What is the median for the data?               8. _____

9. What is the mode for the data?                 9. _____

# Vocabulary Test

**Choose the correct term to complete each sentence.**

| | | |
|---|---|---|
| average | measures of center | outlier |
| first quartile | measures of variation | quartiles |
| mean | median | range |
| mean absolute deviation | mode | third quartile |

1. The _____ is the middle number, or the mean of the two middle numbers, of the ordered data in a set.

   1. _____

2. The _____ of the data set is the sum of the data divided by the number of pieces of data.

   2. _____

3. _____ are values that divide the data set into four equal parts.

   3. _____

4. Extremely high or low values in a data set are called _____.

   4. _____

5. The _____ is the number or numbers that occur most frequently in a data set.

   5. _____

6. The _____ is the difference between the greatest and least data values.

   6. _____

7. _____ is a common term used to refer to the measure of central tendency otherwise known as the mean.

   7. _____

8. Numbers called _____ can be used to describe the center of data.

   8. _____

# Standardized Test Practice

**Read each question. Then fill in the correct answer on the answer sheet provided by your teacher or on a sheet of paper.**

1. A shop records the number of specialty shirts sold each month. What is the mean number of T-shirts sold?

   | T-Shirt Sales | |
   |---|---|
   | **Month** | **Number** |
   | January | 75 |
   | February | 68 |
   | March | 75 |
   | April | 92 |
   | May | 105 |

   **A.** 75  **C.** 85
   **B.** 83  **D.** 92

2. Javon's bill at the bakery was $12. He bought 12 egg bagels for $0.50 each. If an onion bagel costs $0.75, how can he find how much he spent on onion bagels?

   **F.** Add $0.50 and $0.75.
   **G.** Subtract the product of 12 and $0.50 from $12.00.
   **H.** Multiply $0.75 and 12.
   **I.** Divide 12 by $0.50.

3. What is the median of the following ages of people attending a concert: 2, 7, 41, 25, 19, 22, 28, 32, and 24?

   **A.** 17
   **B.** 22
   **C.** 24
   **D.** 41

4. ▨▨ **GRIDDED RESPONSE** Teresa had a family photo with the dimensions of 12 inches by $16\frac{1}{2}$ inches. Find the area of the photo, in square inches.

5. ▨▨ **GRIDDED RESPONSE** Use the frequency table below to determine the mean number of hours of sleep.

   | Hours of Sleep | Tally | Frequency |
   |---|---|---|
   | 5 | II | 2 |
   | 5.5 | I | 1 |
   | 6 | III | 3 |
   | 6.5 | I | 1 |
   | 7 | IIII I | 5 |
   | 7.5 | IIII | 4 |
   | 8 | IIII | 4 |

6. Martha was making place cards for a party that used $\frac{1}{3}$ of a piece of paper each. If she needed to make place cards for 31 people, how many pieces of paper would she need?

   **F.** 93
   **G.** 34
   **H.** 11
   **I.** 10

7. The lengths of the longest bridges in the United States are shown. Which statement is supported by the data?

   | State | Length (ft) |
   |---|---|
   | LA | 126,024 |
   | VA | 79,200 |
   | GA | 42,240 |
   | LA | 96,095 |
   | LA | 120,400 |

   **A.** If the lengths of the bridges were distributed equally among all five bridges, each would measure 95,250 feet in length.

   **B.** There is no bridge length that occurs more often than another.

   **C.** The lengths of the bridges have a range of 82,000 feet.

   **D.** The majority of the bridges are greater than 97,000 feet in length.

8. ▦✎ **GRIDDED RESPONSE** A train is traveling 54.8 miles per hour. At this speed, how far will the train travel, in miles, in 3.2 hours?

9. Craig planted $\frac{3}{4}$ acre with vegetables. He divided the garden into sections that were $\frac{1}{6}$ of an acre. How many different sections will Craig have?

   **F.** $4\frac{1}{2}$

   **G.** $3\frac{1}{2}$

   **H.** $\frac{2}{9}$

   **I.** $\frac{1}{8}$

10. ▦ **SHORT RESPONSE** The table shows students' favorite track and field events. If 300 students were surveyed, how many would prefer long jump?

| Favorite Track and Field Event | |
| --- | --- |
| Discus | 24% |
| Sprints | 31% |
| Distance | 19% |
| Long jump | 26% |

11. The weight of a small dog is about 10 times the weight of a guinea pig shown. What is the estimated weight of the small dog?

2.0 lb

   **A.** 0.2 lb
   **B.** 10 lb
   **C.** 20 lb
   **D.** 200 lb

12. The table shows the number of goals scored by Major League Soccer players in a recent year.

| Name | Goals |
| --- | --- |
| Luciano Emilio | 47 |
| Juan Pablo Angel | 45 |
| Eddie Johnson | 39 |
| Joseph Ngwenya | 37 |
| Juan Toja | 28 |
| Brian Ching | 24 |
| Yura Movsisyan | 22 |
| Taylor Twellman | 20 |

Which measure of center best represents the data?

   **F.** outlier
   **G.** mode
   **H.** range
   **I.** mean

13. What is the solution to the equation $1.7t = 8.5$?

   **A.** 5
   **B.** 6.8
   **C.** 10.2
   **D.** 14.45

14. ▦ **EXTENDED RESPONSE** Kamilah is researching the number of channels available from different cable and satellite companies. The table shows the results.

| Company | Number of Channels |
| --- | --- |
| A | 265 |
| B | 70 |
| C | 85 |
| D | 120 |
| E | 135 |

   **Part A** Find the mean, median, mode, and range.

   **Part B** How does the outlier affect the values you found for Part A?

# Student Recording Sheet

SCORE _____

*Use this recording sheet with the Standardized Test Practice.*

**Fill in the correct answer. For gridded-response questions, write your answers in the boxes on the answer grid and fill in the bubbles to match your answers.**

1. Ⓐ Ⓑ Ⓒ Ⓓ

2. Ⓕ Ⓖ Ⓗ Ⓘ

3. Ⓐ Ⓑ Ⓒ Ⓓ

4.

5.

6. Ⓕ Ⓖ Ⓗ Ⓘ

7. Ⓐ Ⓑ Ⓒ Ⓓ

8.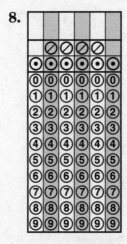

9. Ⓕ Ⓖ Ⓗ Ⓘ

10. _____

11. Ⓐ Ⓑ Ⓒ Ⓓ

12. Ⓕ Ⓖ Ⓗ Ⓘ

13. Ⓐ Ⓑ Ⓒ Ⓓ

## Extended Response

**Record your answers for Exercise 14 on the back of this paper.**

# Extended-Response Test

**Demonstrate your knowledge by giving a clear, concise solution to each problem. Be sure to include all relevant drawings and justify your answers. You may show your solution in more than one way or investigate beyond the requirements of the problem. If necessary, record your answer on another piece of paper.**

1. David has kept track of his family's grocery bills for the past 10 weeks, as shown in the table.

| Week | 1 | 2 | 3 | 4 | 5 | 6 | 7 | 8 | 9 | 10 |
|---|---|---|---|---|---|---|---|---|---|---|
| Bill ($) | 92 | 106 | 129 | 115 | 100 | 84 | 110 | 156 | 98 | 87 |

   **a.** Explain how to find the mean of the data. Then find the mean.

   **b.** Explain how to find the median of the data. Then find the median.

   **c.** Explain why the data has no mode.

   **d.** Explain how to find the range of the data. Then find the range.

   **e.** Find the third and first quartiles and the interquartile range of the data.

   **f.** Explain how to identify outliers in a set of data. Then identify any outliers in the data above.

2. Explain when it is most appropriate to use the mean to describe a data set, the median to describe a data set, and the mode to describe a data set.

# Extended-Response Rubric

| Score | Description |
|:---:|:---|
| 4 | A score of four is a response in which the student demonstrates a thorough understanding of the mathematics concepts and/or procedures embodied in the task. The student has responded correctly to the task, used mathematically sound procedures, and provided clear and complete explanations and interpretations.<br><br>The response may contain minor flaws that do not detract from the demonstration of a thorough understanding. |
| 3 | A score of three is a response in which the student demonstrates an understanding of the mathematics concepts and/or procedures embodied in the task. The student's response to the task is essentially correct with the mathematical procedures used and the explanations and interpretations provided demonstrating an essential but less than thorough understanding.<br><br>The response may contain minor flaws that reflect inattentive execution of mathematical procedures or indications of some misunderstanding of the underlying mathematics concepts and/or procedures. |
| 2 | A score of two indicates that the student has demonstrated only a partial understanding of the mathematics concepts and/or procedures embodied in the task. Although the student may have used the correct approach to obtaining a solution or may have provided a correct solution, the student's work lacks an essential understanding of the underlying mathematical concepts.<br><br>The response contains errors related to misunderstanding important aspects of the task, misuse of mathematical procedures, or faulty interpretations of results. |
| 1 | A score of one indicates that the student has demonstrated a very limited understanding of the mathematics concepts and/or procedures embodied in the task. The student's response is incomplete and exhibits many flaws. Although the student's response has addressed some of the conditions of the task, the student reached an inadequate conclusion and/or provided reasoning that was faulty or incomplete.<br><br>The response exhibits many flaws or may be incomplete. |
| 0 | A score of zero indicates that the student has provided no response at all, or a completely incorrect or uninterpretable response, or demonstrated insufficient understanding of the mathematics concepts and/or procedures embodied in the task. For example, a student may provide some work that is mathematically correct, but the work does not demonstrate even a rudimentary understanding of the primary focus of the task. |

# Test, Form 1A

**Write the correct letter for the correct answer in the blank at the right of each question.**

**For Exercises 1 and 2 use the following set of data:**
**2, 3, 3, 4, 5, 7, 8, 8, 8, 10, 10, and 12.**

1. What are the third and first quartiles of the data?
   **A.** 12, 2     **B.** 8, 4.5     **C.** 10, 3     **D.** 9, 3.5

   1. _____

2. What is the interquartile range of the data?
   **F.** 10            **H.** 7
   **G.** 7.5          **I.** 5.5

   2. _____

3. The table shows the number of movies owned by a group of surveyed people.

| Number of Movies | | | | | |
|----|----|----|----|----|----|
| 11 | 31 | 17 | 22 | 18 | 25 |
| 25 | 10 | 15 | 12 | 30 | 12 |
| 29 | 25 | 21 | 32 | 30 | 25 |

   What is the median of movies owned?
   **A.** 21.5          **C.** 23.5
   **B.** 22            **D.** 25

   3. _____

4. What is the mean absolute deviation of the data: 10, 17, 20, 12, and 16? Round to the nearest whole number.
   **F.** 15           **H.** 3
   **G.** 5            **I.** 2

   4. _____

**For Exercises 5–7, use the dot plot below.**

**Number of Fiction Books Read**

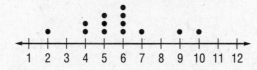

5. What is the mean of the data? Round to the nearest tenth.
   **A.** 5.7       **B.** 5.2       **C.** 5       **D.** 4.5

   5. _____

6. What is the median of the data?
   **F.** 10       **G.** 9       **H.** 5       **I.** 3

   6. _____

7. What is the mode of the data?
   **A.** 3       **B.** 5       **C.** 8       **D.** 10

   7. _____

# Test, Form 1A   *(continued)*

**For Exercises 8–10, which measure of center best represents each set of data?**

8. number of television sets in the home: 4, 4, 2, 4, 4, 1, 3, 4, 2, 4

    **F.** mean        **G.** median        **H.** mode        **I.** range        8. _____

9. goals scored during hockey season: 7, 6, 9, 3, 12, 7, 2, 10, 7, 8

    **A.** mean        **B.** median        **C.** mode        **D.** range        9. _____

10. prices of earrings: $4, $30, $10, $12, $10, $20, $12, $10

    **F.** mean        **G.** median        **H.** mode        **I.** range        10. _____

11. Find the mean absolute deviation of the data set 8, 8, 10, 15, 9, 20, 12, and 10. Round to the nearest hundredth if necessary.

    **A.** 0.52                    **C.** 3.5
    **B.** 3.13                    **D.** 11.5                    11. _____

12. The results of a survey about the number of contacts stored in cell phones is shown in the table.

    | Number of Contacts | | | | | |
    |-----|-----|-----|-----|-----|-----|
    | 100 | 31 | 45 | 85 | 98 | 29 |
    | 75 | 62 | 78 | 50 | 72 | 84 |
    | 50 | 40 | 83 | 88 | 44 | 94 |

    What is the first quartile of the data?

    **F.** 40                    **H.** 45
    **G.** 44.5                    **I.** 85                    12. _____

13. Jamie mowed 7 lawns. He earned $10, $15, $12, $15, $8, and $15 for six lawns. How much did he earn the seventh time if the mean of the data is $12?

    **A.** $9                    **C.** $12
    **B.** $10                    **D.** $15                    13. _____

# Test, Form 1B

**Write the letter for the correct answer in the blank at the right of each question.**

**For Exercises 1 and 2, use the following set of data: 5, 7, 7, 6, 4, 8, 17, 5, 7, 5, 6, and 5.**

1. Find the third and first quartiles of the data.

   **A.** 7, 5      **B.** 6.5, 6      **C.** 17, 4      **D.** 8, 6

   1. _____

2. Find the interquartile range of the data.

   **F.** 2                    **H.** 12

   **G.** 5                    **I.** 17

   2. _____

3. The table shows the prices of sneakers in a store.

   | Price of Sneakers ($) | | | | | |
   |----|----|----|----|----|----|
   | 40 | 37 | 25 | 35 | 29 | 43 |
   | 34 | 26 | 39 | 43 | 51 | 47 |
   | 35 | 27 | 45 | 28 | 50 | 43 |

   What is the median price for the sneakers?

   **A.** $25                  **C.** $38

   **B.** $29                  **D.** $43

   3. _____

4. What is the mean absolute deviation of the data: 10, 20, 15, 17, and 13?

   **F.** 15                   **H.** 3

   **G.** 10                   **I.** 2.8

   4. _____

**For Exercises 5–7, use the dot plot below.**

**Number of Fiction Books Read**

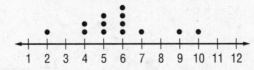

5. What is the mean of the data? Round to the nearest tenth.

   **A.** 13      **B.** 7.5      **C.** 5.8      **D.** 2.9

   5. _____

6. What is the mode of the data?

   **F.** 10      **G.** 6      **H.** 5      **I.** 2

   6. _____

7. What is the median of the data?

   **A.** 4      **B.** 5      **C.** 6      **D.** 7

   7. _____

# Test, Form 1B  (continued)

**For Exercises 8–10, which measure of center best represents each set of data?**

8. runs scored during the baseball season: 8, 6, 9, 3, 12, 4, 2, 10, 7, 8

   **F.** mean          **H.** mode
   **G.** median        **I.** range          8. _____

9. prices of earrings: $6, $12, $15, $12, $12, $20, $12, $13

   **A.** mean          **C.** mode
   **B.** median        **D.** range          9. _____

10. number of telephones in the home: 3, 3, 5, 4, 3, 8, 3, 4, 3, 6, 2

   **F.** mean          **H.** mode
   **G.** median        **I.** range          10. _____

11. Find the mean absolute deviation of the data set 8, 10, 15, 8, 12, 13, 12 and 14. Round to the nearest hundredth if necessary.

   **A.** 1.7           **C.** 2.13
   **B.** 2.05          **D.** 4.25           11. _____

12. The table shows the results of a survey about the number of contacts stored in cell phone.

| Number of Contacts | | | | | |
|---|---|---|---|---|---|
| 100 | 31 | 45 | 85 | 98 | 29 |
| 75 | 62 | 78 | 50 | 72 | 84 |
| 50 | 40 | 83 | 88 | 44 | 94 |

   What is the third quartile of the data?

   **F.** 67            **H.** 84
   **G.** 83            **I.** 85             12. _____

13. Laura mowed 8 lawns. She earned $12, $10, $15, $15, $15, $10, and $9 for seven lawns. How much did she earn the eighth time if the mean of the data is $12?

   **A.** $9            **C.** $12
   **B.** $10           **D.** $15            13. _____

# Test, Form 2A

**Write the letter for the correct answer in the blank at the right of each question.**

**The table shows the number of hours Felisa spent sleeping each night for 12 nights. Use the table to answer Exercises 1–3.**

| Hours Spent Sleeping | | | |
|---|---|---|---|
| 8 | 6 | 7 | 8 |
| 10 | 8 | 8 | 6 |
| 6 | 8 | 8 | 7 |

**1.** What is the mean of the data?

   **A.** 7       **B.** 7.5       **C.** 8       **D.** 8.5           **1.** _____

**2.** What is the median of the data?

   **F.** 7       **G.** 7.5       **H.** 8       **I.** 8.5           **2.** _____

**3.** Which measure of center best represents the data?

   **A.** mean       **B.** median       **C.** mode       **D.** range           **3.** _____

**Refer to the dot plot to answer Exercises 4 and 5.**

**Cookie Dough Tubs
Sold per Student**

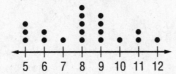

**4.** What is the mode of the data?

   **F.** 7       **G.** 8       **H.** 9       **I.** 10           **4.** _____

**5.** What is the median of the data?

   **A.** 6       **B.** 7       **C.** 8       **D.** 9           **5.** _____

**6.** What is the distance between the first and third quartiles of a data set called?

   **F.** range                **H.** median

   **G.** quartile            **I.** interquartile range           **6.** _____

# Test, Form 2A (continued)

**Use the table for Exercises 7–9.**

| Dress Costs ($) | | | |
|---|---|---|---|
| 38 | 48 | 50 | 38 |
| 52 | 38 | 18 | 48 |

7. What is the mean of the data?

   **A.** $38    **B.** $41.25    **C.** $47.14    **D.** $52

7. _____

8. Which value is the outlier?

   **F.** $52    **G.** $48    **H.** $32    **I.** $18

8. _____

9. What is the mean cost without the outlier? Round to the nearest cent if necessary.

   **A.** $21.50    **B.** $41.25    **C.** $44.57    **D.** $47.14

9. _____

**For Exercises 10–13, use the following set of data: 10, 15, 25, 19, 37, 62, 29, 8, 6, 30, 15, and 20.**

10. What are the third and first quartiles of the data?

10. _____

11. What is the interquartile range of the data?

11. _____

12. Are there any outliers in the data set? Explain.

12. _____

13. Lamar scored a 75, 89, 94, 95, 88, 96, 94, 82, 98, and 96 on his science tests last quarter. What is the mean absolution deviation of the data? Round to the nearest tenth.

13. _____

14. The table shows the number of CDs owned by a group of friends. What is the mean absolute deviation of the data? Round to the nearest tenth.

| Number of CDs | | | | | |
|---|---|---|---|---|---|
| 10 | 6 | 13 | 12 | 4 | 15 |

14. _____

15. Which measure of center would you use to describe a data set with no extreme values and no big gaps? Explain your reasoning.

15. _____

# Test, Form 2B

Write the letter for the correct answer in the blank at the right of each question.

The table shows the number of hours Ladena spent sleeping each night for 12 nights. Use the table to answer Exercises 1–3.

| Hours Spent Sleeping | | | |
|---|---|---|---|
| 8 | 6 | 7 | 7 |
| 10 | 8 | 8 | 10 |
| 8 | 8 | 5 | 7 |

1. What is the mean of the data? Round to the nearest tenth.
   A. 7.5          B. 7.6          C. 7.7          D. 8.2          1. _____

2. What is the mode of the data?
   F. 5          G. 6          H. 7          I. 8          2. _____

3. Which measure of center best represents the data?
   A. mean                    C. mode
   B. median                  D. range          3. _____

Refer to the dot plot to answer Exercises 4 and 5.

### Cookie Dough Tubs
### Sold per Student

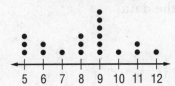

4. What is the median of the data?
   F. 7          G. 8          H. 9          I. 10          4. _____

5. What is the mode of the data?
   A. 5          B. 7          C. 8          D. 9          5. _____

6. What is the median of the data values less than the median called?
   F. third quartile              H. interquartile range
   G. first quartile              I. range          6. _____

# Test, Form 2B    (continued)

Use the table for Exercises 7-9.

| Dress Costs ($) | | | |
|---|---|---|---|
| 38 | 50 | 32 | 42 |
| 50 | 32 | 10 | 50 |

**7.** What is the mean of the data?

  **A.** $38    **B.** $40    **C.** $42    **D.** $50

7. _____

**8.** Which measure of center best describes the data?

  **F.** mean    **G.** median    **H.** mode    **I.** range

8. _____

**9.** What is the mean cost without the outlier? Round to the nearest cent if necessary.

  **A.** $29.40    **B.** $36.75    **C.** $38    **D.** $42

9. _____

**For Exercises 10–13, use the following set of data: 20, 28, 30, 6, 15, 18, 21, 22, 25, 29, 24, and 26.**

**10.** What are the third and first quartiles of the data?

10. _____

**11.** What is the interquartile range of the data?

11. _____

**12.** Are there any outliers in the data set? Explain.

12. _____

**13.** Emma scored a 76, 90, 83, 95, 88, 96, 94, 96, 98, and 96 on her science tests last quarter. What is the mean absolution deviation of the data? Round to the nearest tenth.

13. _____

**14.** The table shows the amount of rainfall in an area for six months. What is the mean absolute deviation of the data?

| Rainfall (in.) | | |
|---|---|---|
| 36 | 28 | 38 |
| 20 | 35 | 89 |

14. _____

# Test, Form 3A

**Write the correct letter for the correct answer in the blank at the right of each question.**

**For Exercises 1 and 2, use the table. It shows the number of books borrowed from a library.**

| Number of Books | | | |
|---|---|---|---|
| 1 | 3 | 15 | 24 |
| 37 | 55 | 39 | 40 |
| 35 | 28 | 20 | 0 |

1. Find the mean, median, and mode of the data.

1. _____

2. _____

2. Which measure of center best describes the data?

**For Exercises 3–6, use the data of test grades shown in the table.**

| Test Grades |
|---|
| 0, 59, 63, 68, 71, 74 |
| 82, 83, 85, 88, 90 |

3. Find the third and first quartiles of the data.

3. _____

4. Find the interquartile range of the data.

4. _____

5. Are there any outliers? Explain.

5. _____

6. Jordan had the following scores on her math tests last quarter: 96, 89, 79, 85, 87, 94, 96, and 98. Find the mean absolute deviation for the set of data. Round to the nearest tenth. How many data values are closer than one mean absolute deviation away from the mean?

6. _____

**For Exercises 7 and 8, use the table. The table shows the results of class survey about their length at birth.**

| Length at Birth (inches) | |
|---|---|
| 19 | 17 |
| 21 | 20 |
| 22 | 14 |
| 18 | 16 |
| 19 | 20 |
| 15 | 19 |
| 14 | 18 |

7. What is the mean of the data?

7. _____

8. What is the median and mode of the data?

8. _____

# Test, Form 3A  (continued)

SCORE _____

**Refer to the table of book costs for Exercises 9–12.**

| Book Costs ($) | | | | |
|---|---|---|---|---|
| 5 | 15 | 9 | 25 | 5 |
| 42 | 11 | 15 | 5 | 18 |

9. Identify the outlier.

9. _____

10. Tell which measure of center best describes the data with and without the outlier.

10. _____

11. Determine how the outlier affects the mean, median, mode, and range of the data.

11. _____

12. What is the mean absolute deviation of the data? Round to the nearest tenth.

12. _____

13. What is the mean absolute deviation of the data: 15.6, 19.8, 12.2, 21.7, 25.3, 16.5, and 19.8? Round to the nearest tenth.

13. _____

**Refer to the table of test scores for Exercises 14–15.**

| Test Scores | | | |
|---|---|---|---|
| 68 | 73 | 72 | 79 |
| 81 | 79 | 91 | 79 |
| 85 | 92 | 95 | 87 |
| 66 | 87 | 96 | 90 |

14. Find the mean of the data.

14. _____

15. What is the median and mode of the data?

15. _____

16. Tell which measure of center best describes the data. Justify your response.

16. _____

# Test, Form 3B

**Write the correct letter for the correct answer in the blank at the right of each question.**

**For Exercises 1 and 2, use the table. It shows the number of students in each extracurricular school club.**

| Number of Students | | | |
|----|----|----|----|
| 10 | 12 | 15 | 73 |
| 13 | 20 | 12 | 16 |
| 15 | 25 | 9  | 14 |

1. Find the mean, median, and mode of the data.

1. _____

2. Which measure of center best describes the data?

2. _____

**For Exercises 3–6, use the data shown in the table.**

| Number of CDs | | | | | |
|----|----|----|----|----|----|
| 10 | 6  | 13 | 12 | 0  | 15 |
| 12 | 0  | 15 | 9  | 11 | 21 |
| 11 | 1  | 14 | 9  | 15 | 22 |

3. Find the third and first quartiles of the data.

3. _____

4. Find the interquartile range of the data.

4. _____

5. Are there any outliers? Explain.

5. _____

6. Serge had the following scores on his math tests last quarter: 100, 94, 78, 88, 87, 93, 96, and 92. Find the mean absolute deviation for the set of data. Round to the nearest tenth. How many data values are closer than one mean absolute deviation away from the mean?

6. _____

**For Exercises 7 and 8, refer to the table. The table shows the birth weights of babies born in a hospital.**

| Weight at Birth (pounds) | |
|------|------|
| 6.2  | 6.5  |
| 5.5  | 6.6  |
| 6.0  | 7.2  |
| 8.4  | 5.0  |
| 9.5  | 7.4  |
| 7.4  | 7.5  |
| 9.0  | 8.2  |

7. What is the mean of the data? Round to the nearest tenth.

7. _____

8. What is the median and mode of the data?

8. _____

# Test, Form 3B  *(continued)*

**Refer to the table of book costs for Exercises 9–12.**

| Book Costs ($) | | |
|---|---|---|
| 1 | 1 | 7 |
| 18 | 1 | 5 |

**9.** Identify the outlier.

9. _____

**10.** Determine how the outlier affects the mean, median, mode, and range of the data.

10. _____

**11.** Tell which measure of center best describes the data with and without the outlier.

11. _____

**12.** What is the mean absolute deviation of the data? Round to the nearest tenth.

12. _____

**13.** What is the mean absolute deviation of the data: 17.9, 19.3, 18.8, 20.5, 22.6, 17.4, and 19.8? Round to the nearest tenth.

13. _____

**Refer to the table of test scores for Exercises 14–15.**

| Test Scores | | | |
|---|---|---|---|
| 69 | 95 | 92 | 87 |
| 72 | 87 | 84 | 78 |
| 82 | 80 | 68 | 78 |
| 89 | 78 | 98 | 91 |

**14.** Find the mean of the data.

14. _____

**15.** What is the median and mode of the data?

15. _____

**16.** Tell which measure of center best describes the data. Justify your response.

16. _____

# Are You Ready?

## Review

> To find the average or **mean** of a set of data, find the sum of the numbers. Then divide the sum by how many numbers are in the set. The resulting number is the mean of the data in the set.

## Example 1

**Find the mean for the data set.**

20, 21, 13, 9, 12

To find the mean, find the sum of the numbers. Then divide the sum by how many numbers are in the set.

$$\frac{20 + 21 + 13 + 9 + 12}{5} = \frac{75}{5}$$
$$= 15$$

## Example 2

**BASKETBALL** Paula entered the local 4-day free-throw tournament. The place she was in each day was 14$^{th}$, 6$^{th}$, 7$^{th}$, and 13$^{th}$. What was the mean of the data set?

$$\frac{14 + 6 + 7 + 13}{4} = \frac{40}{4}$$
$$= 10$$

## Exercises

**Find the mean for each data set. Round to the nearest tenth if necessary.**

1. 18, 16, 19, 13, 24

1. _____

2. 20, 20, 32, 18, 45, 39

2. _____

3. 18, 39, 15, 42, 37, 52, 19, 48

3. _____

4. 62, 21, 18, 23

4. _____

5. **GYMNASTICS** Tommy competes in the pommel horse event. His scores in the last four meets were 7.1, 8.9, 9.5, and 8.5. What is his mean score?

5. _____

# Are You Ready?

## Practice

**Find the median of each set of data.**

1.

| Top Ten Fastest Mammals on Land | |
| --- | --- |
| **Animal** | **Speed (mph)** |
| Cheetah | 71 |
| Pronghorn Antelope | 57 |
| Blue Wildebeest | 50 |
| Lion | 50 |
| Springbok | 50 |
| Brown Hare | 48 |
| Red Fox | 48 |
| Grant's Gazelle | 47 |
| Thomson's Gazelle | 47 |
| Horse | 45 |

2.

| Maria's English Test Scores | | | |
| --- | --- | --- | --- |
| 65 | 90 | 88 | 65 |
| 82 | 78 | 92 | 96 |

**Find the mean (average) for each data set.**
**Round to the nearest tenth if necessary.**

3. 9, 10, 5, 8, 12, 4, 6, 10

4. 10, 29, 19, 22, 18, 10

5. 46, 27, 13, 33, 18, 32, 46, 15

6. 49, 19, 34, 17, 24, 32, 52, 49

7. **BOWLING** Amy's bowling scores at this spring's
bowl-a-thon were 132, 125, 155, and 100. What
was her average score?

1. _____

2. _____

3. _____

4. _____

5. _____

6. _____

7. _____

# Are You Ready?

## Apply

1. **GAMES** Jewel likes to play video games. The table shows her scores. What was her average score? Round to the nearest tenth if necessary.

| Jewel's Scores | | |
|---|---|---|
| 78 | 69 | 72 |
| 71 | 71 | 69 |
| 78 | 72 | 52 |

2. **SPRING** Karly owns a lawn service. During the month of May she averaged 47 customers the first week, 38 the second week, 55 the third week, and 60 the fourth week. How many customers did she average per week?

3. **CANDY** The number of bags of candy sold at a certain movie theater for several days was 32, 17, 20, 38, 47, and 20. Find the average number of bags of popcorn sold during these days.

4. **CUSTOMERS** A store keeps track of the number of customers entering the store in the first hour of each day. In the first four days of June, the store had 215, 143, 100, and 214 customers in the first hour. Find the average number of customers the store had in the first hour during those four days.

5. **KENNELS** Betsy trains dogs. The table shows how many dogs she trained during the first six months this year. What is the median of the data?

| Number of Dogs Trained | | |
|---|---|---|
| 21 | 11 | 18 |
| 23 | 30 | 15 |

6. **CHOIR** Timmy belongs to the municipal choir in his home town. The table below shows the age of each member of the choir. What is the median of the data?

| Choir Member Ages | | | | | |
|---|---|---|---|---|---|
| 18 | 25 | 28 | 33 | 35 | 42 |
| 24 | 19 | 35 | 47 | 35 | 44 |

# Diagnostic Test

**Find the median for each data set.**

1.

| Monthly Mean Temperatures in Pensacola, Florida | | | |
|----|----|----|----|
| 51 | 54 | 54 | 60 |
| 61 | 68 | 68 | 75 |
| 78 | 80 | 82 | 82 |

1. _____

2.

| Number of Canaries Raised | | |
|----|----|----|
| 42 | 33 | 41 |
| 45 | 30 | 29 |

2. _____

**Find the mean (average) for each data set. Round to the nearest tenth if necessary.**

3. 14, 45, 31, 30

3. _____

4. 20, 75, 77, 11, 23, 22

4. _____

5. 15, 27, 11, 19, 31, 19, 25, 5

5. _____

6. 52, 20, 34, 12, 8, 14

6. _____

7. 13, 27, 15, 63, 12, 9, 32, 90

7. _____

8. 53, 40, 31, 30, 32, 56, 51

8. _____

9. **TOURISTS** Clarine owns a sight-seeing boat in Florida. In each of the last five weeks she had 57, 59, 67, 51, and 76 customers. How many customers did she average per week?

9. _____

# Pretest

**For 1-3, use the table. The table shows the top speeds of sixteen roller coasters.**

| Top Speeds of Roller Coasters (mph) | | | |
|---|---|---|---|
| 80 | 40 | 48 | 70 |
| 93 | 95 | 60 | 40 |
| 57 | 40 | 65 | 90 |
| 92 | 72 | 60 | 72 |

1. Make a box plot of the data.

1. _____

2. What percent of roller coasters had speeds 85 miles per hour or more?

2. _____

3. Find the median and the measures of variability.

3. _____

**For Exercises 4–6, refer to the histogram that shows the number of attendees at a festival according to various age intervals.**

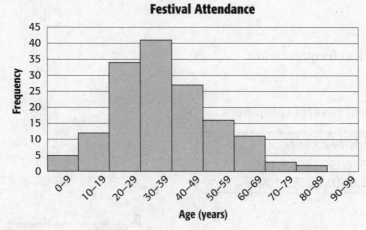

**Festival Attendance**

4. Which age interval had the greatest number of attendees?

4. _____

5. Which age interval had more attendees: 20–29 or 40–49?

5. _____

6. Which age interval had fewer attendees: 70–79 or 0–9?

6. _____

# Chapter Quiz

**For Exercises 1 and 2 use the line plot.**

1. Use the line plot. What is the mode of the data?

**Math Grades**

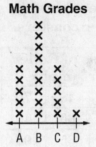

1. _____

2. Compare the number of students who scored a B to the number who scored an A.

2. _____

**EARTHQUAKES For Exercises 3 and 4, use the table.**

| Major Earthquakes | |
|---|---|
| **Magnitude** | **Frequency** |
| 7.0–7.3 | 7 |
| 7.4–7.7 | 5 |
| 7.8–8.1 | 1 |
| 8.2–8.5 | 1 |

3. Describe the data in the table.

3. _____

4. Draw a histogram to represent the frequency of each interval.

**COMPUTER** The table shows the cost of several computers.

| Cost of Computer ($) |
|---|
| 500, 950, 1,120, 1,480, 699, 780, 2,450, 2,800, 875, 565, 820, 1,000, 1,299, 1,325 |

4. _____

5. Find the range of the data.

5. _____

6. Find the third and first quartiles.

6. _____

7. Find any outliers of the data.

7. _____

**Cost of Computers**

8. Draw a box plot to show this data.

8.

# Vocabulary Test

| | | |
|---|---|---|
| box plot | frequency distribution | line plot |
| cluster | gap | peak |
| distribution | histogram | symmetric |
| dot plot | line graph | |

## Choose a word from the list to complete each sentence.

1. A(n) _____ is a display that shows the distribution data where the data is represented by an X.

1. _____

2. A _____ is an empty space or interval in a set of data.

2. _____

3. Data that are grouped together are called a _____.

3. _____

4. The _____ of a set of data shows the arrangement of data.

4. _____

5. A _____ is a diagram that is constructed using the median, quartiles, and extreme values.

5. _____

6. The _____ is the most frequently occuring value, or mode of the data plot.

6. _____

7. A _____ is a type of bar graph that displays numerical data that have been organized into equal intervals.

7. _____

## Define each term in your own words.

8. _____

8. symmetric

9. _____

9. frequency distribution

# Standardized Test Practice

**Read each question. Then fill in the correct answer on the answer sheet provided by your teacher or on a sheet of paper.**

1. Which of the following is the best to display data that is divided into equal intervals?
   A. bar graph
   B. box plot
   C. histogram
   D. line plot

2. The table shows the number of wins the Bears baseball team had in the last 4 seasons. What is the mean number of wins?

   | Year | 2007 | 2008 | 2009 | 2010 |
   |------|------|------|------|------|
   | Wins | 6 | 7 | 13 | 10 |

   F. 6
   G. 9
   H. 10
   I. 13

3. Titus ate $\frac{3}{8}$ of a lasagna. If the lasagna was divided into 16 equal sections, how many sections did Titus eat?
   A. 6
   B. 8
   C. 10
   D. 16

4. **SHORT RESPONSE** Ethan and two friends are going to split the cost of a video game equally. How much will each person pay if the cost of the video game is $48.36?

5. **GRIDDED RESPONSE** What is mode of the data set?

   number of cars: 12, 15, 17, 18, 12

6. The circle graph shows the percentages of the colors of cars purchased at a car dealership. What fraction of the cars purchased are red?

   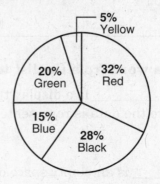

   F. $\frac{1}{2}$     H. $\frac{17}{50}$

   G. $\frac{47}{100}$     I. $\frac{8}{25}$

7. **SHORT RESPONSE** What is the area of the triangle below?

   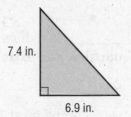

8. Malik was estimating the area of his room to replace the carpet. Which measurement would be a reasonable estimate of the area of his room?

   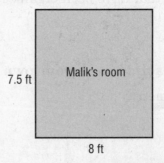

   A. 15 square feet     C. 64 square feet
   B. 20 square feet     D. 72 square feet

9. 📝 **GRIDDED RESPONSE** Six out of every 25 students usually receive an A on the history test. Predict the number of students out of 125 that received an A on the history test.

10. The histogram shows the number of yards gained on first downs. How many times did the team gain at least 6 yards on first downs?

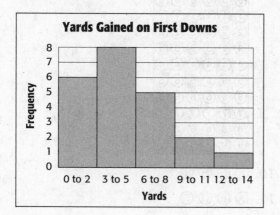

**Yards Gained on First Downs**

F. 9

G. 8

H. 7

I. 4

11. Which of the following would you use to describe the center of a data set, if the data set is symmetric?

A. median

B. mode

C. range

D. mean

12. 📝 **SHORT RESPONSE** The volume of a rectangular prism is 336 cubic centimeters. The prism has a length of 8 centimeters and a width of 6 centimeters. What is the height of the prism?

13. On a field trip, there is one adult for every 8 students. There are 136 students on the field trip. Which equivalent ratio could be used to find the number of adults?

| Students | Adults |
|----------|--------|
| 8 | 1 |
| 16 | 2 |
| 136 | ■ |

F. $\dfrac{■}{8} = \dfrac{1}{136}$

G. $\dfrac{1}{8} = \dfrac{■}{136}$

H. $\dfrac{8}{1} = \dfrac{■}{136}$

I. $\dfrac{■}{1} = \dfrac{8}{136}$

14. 📝 **EXTENDED RESPONSE** Kamilah is researching the number of channels available from different cable and satellite companies. The table shows the results.

| Company | Number of Channels |
|---------|--------------------|
| A | 265 |
| B | 70 |
| C | 85 |
| D | 120 |
| E | 135 |

**Part A** Select and make an appropriate display that best represents the data.

**Part B** Find the mean, median, mode, and range.

**Part C** How does the outlier affect the values you found for Part B?

# Student Recording Sheet

*Use this recording sheet with the Standardized Test Practice pages.*

**Fill in the correct answer. For gridded-response questions, write your answers in the boxes on the answer grid and fill in the bubbles to match your answers.**

1. Ⓐ Ⓑ Ⓒ Ⓓ

2. Ⓕ Ⓖ Ⓗ Ⓘ

3. Ⓐ Ⓑ Ⓒ Ⓓ

4. _____

5.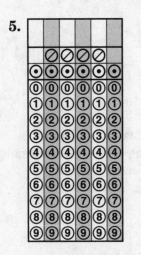

6. Ⓕ Ⓖ Ⓗ Ⓘ

7. _____

8. Ⓐ Ⓑ Ⓒ Ⓓ

9.

10. Ⓕ Ⓖ Ⓗ Ⓘ

11. Ⓐ Ⓑ Ⓒ Ⓓ

12. _____

13. Ⓕ Ⓖ Ⓗ Ⓘ

## Extended Response

**Record your answers for Exercise 14 on the back of this paper.**

# Extended-Response Test

Demonstrate your knowledge by giving a clear, concise solution to each problem. Be sure to include all relevant drawings and justify your answers. You may show your solution in more than one way or investigate beyond the requirements of the problem. If necessary, record your answer on another piece of paper.

1. Make a frequency table of the data in the table about student grades.

| Students' Grades |   |   |   |   |   |   |   |
|---|---|---|---|---|---|---|---|
| A | C | C | B | C | A | C | B | F |
| B | D | A | C | D | C | D | C | F |
| C | C | B | B | A | B | C | D |   |

2. Use your frequency table from Exercise 1.
   a. Select and make the appropriate display for the data. Be sure to label all the parts of the graph.

   b. Explain how you decided on the scale and interval.

   c. Describe what your graph shows.

   d. How does the number of students with a grade of A compare with the number of students with a grade of D? with a grade of F?

3. Use the test scores at the right.
   a. Make a histogram of the data in the table of test scores.

   | Scores on a Test |   |   |   |   |   |
   |---|---|---|---|---|---|
   | 94 | 72 | 82 | 82 | 86 | 66 |
   | 76 | 34 | 98 | 90 | 78 | 92 |
   | 68 | 82 | 76 | 76 | 78 | 84 |
   | 78 | 88 | 84 | 94 | 82 | 80 |

   b. Write a sentence that analyzes the data from the graph.

   c. Find the mean, median, mode, and range of the data in the table. Show your work.

   d. Use the table. Identify any outliers and find the mean without the outliers. Describe how the outlier affects the mean of the data.

# Extended-Response Rubric

| Score | Description |
|:---:|:---|
| 4 | A score of four is a response in which the student demonstrates a thorough understanding of the mathematics concepts and/or procedures embodied in the task. The student has responded correctly to the task, used mathematically sound procedures, and provided clear and complete explanations and interpretations.<br><br>The response may contain minor flaws that do not detract from the demonstration of a thorough understanding. |
| 3 | A score of three is a response in which the student demonstrates an understanding of the mathematics concepts and/or procedures embodied in the task. The student's response to the task is essentially correct with the mathematical procedures used and the explanations and interpretations provided demonstrating an essential but less than thorough understanding.<br><br>The response may contain minor flaws that reflect inattentive execution of mathematical procedures or indications of some misunderstanding of the underlying mathematics concepts and/or procedures. |
| 2 | A score of two indicates that the student has demonstrated only a partial understanding of the mathematics concepts and/or procedures embodied in the task. Although the student may have used the correct approach to obtaining a solution or may have provided a correct solution, the student's work lacks an essential understanding of the underlying mathematical concepts.<br><br>The response contains errors related to misunderstanding important aspects of the task, misuse of mathematical procedures, or faulty interpretations of results. |
| 1 | A score of one indicates that the student has demonstrated a very limited understanding of the mathematics concepts and/or procedures embodied in the task. The student's response is incomplete and exhibits many flaws. Although the student's response has addressed some of the conditions of the task, the student reached an inadequate conclusion and/or provided reasoning that was faulty or incomplete.<br><br>The response exhibits many flaws or may be incomplete. |
| 0 | A score of zero indicates that the student has provided no response at all, or a completely incorrect or uninterpretable response, or demonstrated insufficient understanding of the mathematics concepts and/or procedures embodied in the task. For example, a student may provide some work that is mathematically correct, but the work does not demonstrate even a rudimentary understanding of the primary focus of the task. |

NAME _____ DATE _____ PERIOD _____

# Test, Form 1A

SCORE _____

**Write the letter for the correct answer in the blank at the right of each question.**

1. Which of the following is an appropriate display to show the height of a plant over the past 3 weeks?
   - **A.** bar graph
   - **B.** line graph
   - **C.** circle graph
   - **D.** histogram

   1. _____

2. Which of the following is an appropriate display to show the heights of sixth graders arranged by intervals?
   - **F.** bar graph
   - **G.** line graph
   - **H.** circle graph
   - **I.** histogram

   2. _____

**For Exercises 3–4, use the box plot. It shows the number of days on the market for single family homes in a city.**

**Home Sales: Days on the Market**

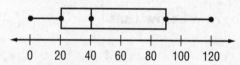

3. What are the third and first quartiles of the data?
   - **A.** 60, 0
   - **B.** 40, 20
   - **C.** 90, 20
   - **D.** 120, 0

   3. _____

4. What percent of homes were on the market between 20 and 40 days?
   - **F.** 25%
   - **G.** 50%
   - **H.** 75%
   - **I.** 100%

   4. _____

**For Exercises 5–7 use the following line plot. It shows the number of cans collected by the student council.**

**Number of Cans Collected**

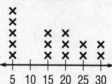

5. What is the mean of the data?
   - **A.** 5
   - **B.** 15
   - **C.** 16
   - **D.** 20

   5. _____

6. Which of the following describes the shape of the data distribution?
   - **F.** It is symmetric.
   - **G.** There is a peak at 5.
   - **H.** There is a cluster from 5–10.
   - **I.** There is a gap from 15–30.

   6. _____

7. Which is best used to describe the center of the data?
   - **A.** mean
   - **B.** median
   - **C.** mode
   - **D.** range

   7. _____

# Test, Form 1A     *(continued)*

**For Exercises 8 and 9, use the line plot that shows the number of times students went to the pool in June.**

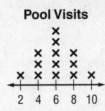

**Pool Visits**

8. Which of the following describes the data?

   **F.** symmetric, peak at 6      **H.** symmetric, peak at 2

   **G.** not symmetric               **I.** symmetric, peak at 8          8. _____

9. Which best describes the center of the data?

   **A.** 2                           **C.** 6

   **B.** 4                           **D.** 8                              9. _____

**For Exercises 10 and 11, refer to the histogram below that shows the prices of cameras.**

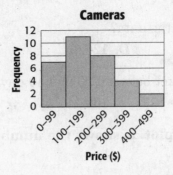

**Cameras**

10. What is the cost of the greatest number of cameras?

    **F.** $11                        **H.** $100–$199

    **G.** $0–$99                     **I.** $400–$499                     10. _____

11. How many cameras cost more than $299?

    **A.** 6                          **C.** 14

    **B.** 8                          **D.** 18                            11. _____

12. The graph shows the number of students who cleared the bar on the pole vault at various heights. Which of the following is the best prediction for the number of students who would clear 15 feet?

    **F.** 25                         **H.** 12

    **G.** 14                         **I.** 5                             12. _____

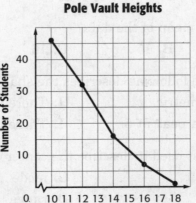

**Pole Vault Heights**

# Test, Form 1B

**Write the letter for the correct answer in the blank at the right of each question.**

1. Which of the following is an appropriate display to show interest rates over the past 3 weeks?

   **A.** bar graph   **B.** line graph   **C.** circle graph   **D.** histogram

   1. _____

2. Which of the following is an appropriate display to show the heights of adults arranged by intervals?

   **F.** bar graph   **G.** line graph   **H.** circle graph   **I.** histogram

   2. _____

**For Exercises 3–4, use the box plot. It shows the number of days on the market for single family homes in a city.**

**Days to Sell Homes**

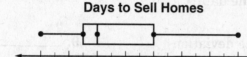

3. What are the third and first quartiles of the data?

   **A.** 40, 30          **C.** 140, 0

   **B.** 140, 80         **D.** 80, 30

   3. _____

4. What percent of homes were on the market between 40 and 140 days?

   **F.** 25%        **G.** 50%        **H.** 75%        **I.** 100%

   4. _____

**For Exercises 5–7, use the line plot below. It shows the number of cans collected by the student council.**

**Number of Cans Collected**

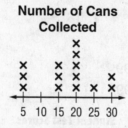

5. What is the mean of the data?

   **A.** 15          **B.** 17.5          **C.** 22.5          **D.** 25

   5. _____

6. Which of the following describes the shape of the data distribution?

   **F.** There is a peak at 15.      **H.** There is a gap from 25–30.

   **G.** There is a cluster from 5–15.  **I.** It is not symmetric.

   6. _____

7. Which of the following describes the data?

   **A.** symmetric          **C.** peak at 15

   **B.** not symmetric          **D.** cluster at 10

   7. _____

# Test, Form 1B  (continued)

**For Exercises 8 and 9, use the line plot that shows the number of times students went to the pool in June.**

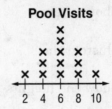

**Pool Visits**

8. Which of the following is true?
   F. the data is not symmetric
   G. peak at 4
   H. no gaps
   I. no mode

8. _____

9. Which would you use to describe the spread of the data?
   A. clusters
   B. interquartile range
   C. range
   I. mean absolute deviation

9. _____

**For Exercises 10 and 11, refer to the histogram.**

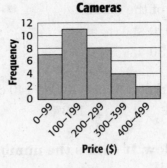

**Cameras**

10. How many cameras cost less than $100?
    F. 2          G. 7          H. 8          I. 18

10. _____

11. Which price range has the least frequency?
    A. $0–$99
    B. $300–$399
    C. $100–$199
    D. $400–$499

11. _____

12. The graph shows test scores of students with various grade point averages. What is the best prediction of a student with a grade point average of 3.25?

    F. 34
    G. 32
    H. 29
    I. 25

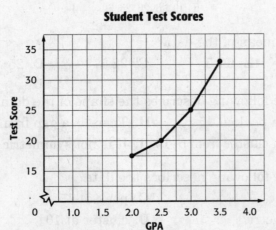

**Student Test Scores**

12. _____

# Test, Form 2A

**Write the letter for the correct answer in the blank at the right of each question.**

**For Exercises 1–4, use the following line plot that shows the number of jumping jacks students completed in 30 seconds.**

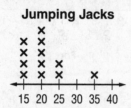

**Jumping Jacks**

1. What is the median of the data?
   A. 15                    C. 25
   B. 20                    D. 35

   1. _____

2. What is the mode?
   F. 15          G. 20          H. 25          I. 35

   2. _____

3. What is the peak of the data?
   A. 15                    C. 25
   B. 20                    D. 35

   3. _____

4. Which of the following best describes the data?
   F. symmetric              H. gap at 15–25
   G. not symmetric          I. range of 10

   4. _____

5. Which of the following is an appropriate display to show the scores on an eye test arranged in intervals?
   A. bar graph              C. circle graph
   B. line graph             D. histogram

   5. _____

**For Exercises 6 and 7, use the box plot. It shows the number of days on the market for single family homes in a city.**

**Home Sales: Days on the Market**

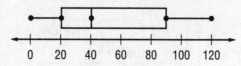

6. What is the interquartile range of the data?
   F. 70          G. 40          H. 90          I. 120

   6. _____

7. What percent of the homes were on the market more than 40 days?
   A. 25%          B. 50%          C. 75%          D. 100%

   7. _____

# Test, Form 2A   (continued)

**For Exercises 8 and 9, refer to the histogram.**

8. Which price range has the least frequency?

   F. $2             H. $20–$39

   G. $0–$19         I. $80–$99

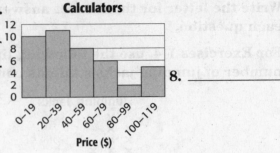

**Calculators**

8. _____

9. How many calculators cost less than $60?

   A. 13             C. 29

   B. 21             D. 35

9. _____

**The line graph shows the number of greeting cards, in thousands, created over several years.**

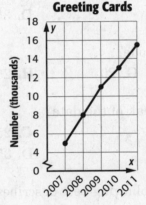

**Greeting Cards**

10. If the trend continues, what is the best prediction for the number of cards that will be created in 2014?

    F. 23,000

    G. 18,000

    H. 23

    I. 18

10. _____

11. The lengths in inches of fish that Ricky caught on a fishing trip are shown below. Find the measures of variation for the data set.

    9.2, 8.7, 8.4, 8.9, 9.5, 9.3, 8.8, 8.9, 8.8, 8.2, 9.6, 9.1

11. _____

12. Draw a box plot for the data in Exercise 11.

**12.**
**Length of Fish (in.)**

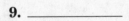

13. Jabez surveyed people about their favorite sports. The results are shown in the table. Which type of display would you use to show the results?

| Sport | Students |
|-------|----------|
| Basketball | 25 |
| Football | 28 |
| Golf | 15 |
| Volleyball | 32 |

13. _____

# Test, Form 2B

SCORE _____

Write the letter for the correct answer in the blank at the right of each question.

**For Exercises 1–4, use the following line plot that shows the number of jumping jacks students completed in 30 seconds.**

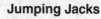

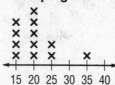

Jumping Jacks

15 20 25 30 35 40

**1.** What is the mean of the data? Round to the nearest tenth.

   **A.** 20.4      **B.** 20      **C.** 17.5      **D.** 15           1. _____

**2.** What is a gap in the data?

   **F.** 15      **G.** 20      **H.** 25      **I.** 30           2. _____

**3.** What would you use to describe the center of the data?

   **A.** mean      **B.** median      **C.** mode      **D.** range           3. _____

**4.** What would you use to describe the spread of the data?

   **F.** interquartile range      **H.** gap

   **G.** range      **I.** mean absolute deviation      4. _____

**5.** Which of the following is an appropriate display to show the average price of a postage stamp over the last 20 years?

   **A.** line graph      **C.** circle graph

   **B.** bar graph      **D.** histogram      5. _____

**For Exercises 6 and 7, use the box plot. It shows the number of days on the market for single family homes in a city.**

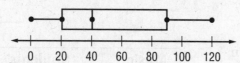

Home Sales: Days on the Market

0 20 40 60 80 100 120

**6.** What is the median of the data?

   **F.** 30      **G.** 40      **H.** 90      **I.** 120           6. _____

**7.** What percent of the homes were on the market less than 90 days?

   **A.** 0%      **B.** 25%      **C.** 50%      **D.** 75%           7. _____

# Test, Form 2B  (continued)

**For Exercises 8 and 9, refer to the histogram. It shows the prices of different calculators.**

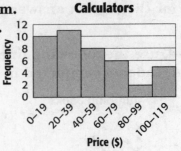

**Calculators**

8. Which price range has the greatest frequency?

   **F.** $100–$119      **H.** $80–$99

   **G.** $20–$39      **I.** $11

8. _____

9. How many calculators cost $80 or more?

   **A.** 7      **B.** 6      **C.** 3      **D.** 2

9. _____

**The line graph shows the number of band members in a high school for several years.**

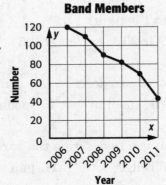

**Band Members**

10. If the trend continues, what is the best prediction for the number of students that will be band members in 2012?

   **F.** 25      **H.** 59

   **G.** 45      **I.** 100

10. _____

11. The number of toys donated by students in 12 classes is shown below. Find the measures of variation for the data set.

    16, 16, 17, 19, 20, 23, 24, 25, 29, 31, 33, 38

11. _____

12. Draw a box plot for the data in Exercise 11.

**12.**
**Number of Toys**

       15  20  25  30  35  40

13. Jonez surveyed people about their favorite music. The results are shown in the table. Which type of display would be best to show the survey results?

| Music | Students |
|---|---|
| Classical | 25 |
| Country | 30 |
| Rock | 30 |
| Jazz | 15 |

13. _____

# Test, Form 3A

Write the letter for the correct answer in the blank at the right of each question.

For Exercises 1–6, use the box plot. It shows the test scores of a history class.

**Test Scores**

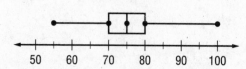

1. What is the interquartile range of the data?

1. _____

2. What is the third and first quartiles of the data?

2. _____

3. What percent of students scored at least 70 on the test?

3. _____

4. Is the data symmetric? Explain.

4. _____

5. Which measure would you use to describe the center of the data?

5. _____

6. Which measure would you use to describe the spread of the data?

6. _____

7. The line graph shows the sales, in hundreds of dollars, of magazines from a business for the past several years. If the trend continues, predict what the sales will be in 2015.

7. _____

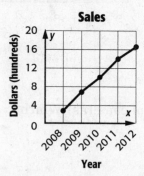

8. The table shows the lengths (in seconds) of student's favorite hit singles. Make a histogram to represent the data.

| Length of Hit Single (s) | | | | |
|------|------|------|------|------|
| 220 | 150 | 220 | 205 | 256 |
| 178 | 261 | 258 | 327 | 275 |
| 166 | 341 | 157 | 208 | 219 |
| 184 | 265 | 225 | 329 | 248 |

8.

# Test, Form 3A    (continued)

**For Exercises 9–11, use the line plot that shows the costs of dresses.**

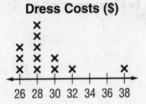

Dress Costs ($)

9. What is the range of the data?

9. _____

10. What is the peak of the data?

10. _____

11. Choose the appropriate measures to describe the center and spread of distribution. Justify your response based on the shape of the distribution.

11. _____

**For Exercises 12 and 13, refer to the histogram. It shows the cost of cameras at a store.**

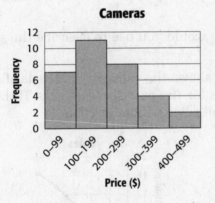

Cameras

12. Which interval has the greatest number of cameras? Which measure of center does this represent?

12. _____

13. How many cameras cost between $200 and $399?

13. _____

14. The table shows the number of pets owned by students in a class. Make a line plot of the data.

| Number of Pets | | | | | |
|---|---|---|---|---|---|
| 0 | 6 | 1 | 2 | 0 | 1 |
| 2 | 0 | 5 | 2 | 1 | 2 |
| 1 | 1 | 4 | 1 | 1 | 2 |

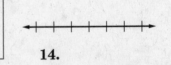

14. _____

# Test, Form 3B

**Write the letter for the correct answer in the blank at the right of each question.**

**For Exercises 1–6, use the box plot. It shows the test scores of an English class.**

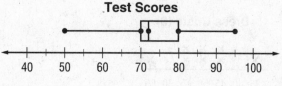

Test Scores

1. What is the interquartile range of the data?

1. _____

2. What is the third and first quartiles of the data?

2. _____

3. About what percent of students scored less than 80 on the test?

3. _____

4. Is the data symmetric? Explain.

4. _____

5. Which measure would you use to describe the center of the data?

5. _____

6. Which measure would you use to describe the spread of the data?

6. _____

7. The line graph shows the bushels of greenbeans, in hundreds, harvested by an Illinois farmer for several years. If the trend continues, predict the number of bushels harvested in 2012.

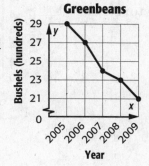

Greenbeans

7. _____

8. The table shows the ages of actors who starred in a series of movies. Make a histogram to represent the data.

| Ages of Actors (yr) | | | | | |
|----|----|----|----|----|----|
| 44 | 10 | 24 | 5  | 29 | 30 |
| 28 | 29 | 18 | 50 | 23 | 3  |
| 24 | 26 | 8  | 34 | 20 | 24 |
| 24 | 25 | 9  | 15 | 39 | 16 |

8.

# Test, Form 3B   (continued)

**For Exercises 9–11, use the line plot that shows the costs of dresses.**

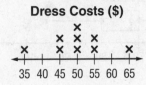

**Dress Costs ($)**

35 40 45 50 55 60 65

9. What is the range of the data?

9. _____

10. What is the peak of the data?

10. _____

11. Choose the appropriate measure to describe the center and spread of distribution. Justify your response based on the shape of the distribution.

11. _____

**For Exercises 12 and 13, refer to the histogram. It shows the cost of calculators at a store.**

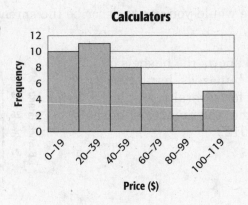

**Calculators**

Frequency

12
10
8
6
4
2
0

0–19   20–39   40–59   60–79   80–99   100–119

**Price ($)**

12. Which price range has the least frequency? Which price range has the greatest frequency?

12. _____

13. How many calculators cost $60 or more?

13. _____

14. The table shows the amount of rainfall in an area over 10 years. Make a line plot of the data.

| Annual Rainfall (in.) | | | | |
|---|---|---|---|---|
| 27 | 28 | 33 | 28 | 30 |
| 30 | 25 | 28 | 27 | 29 |

14.

# Course 1 Benchmark Test – First Quarter (Chapters 1–3)

1. The table below shows the number of pennies, nickels, dimes, and quarters that Heather has in her purse. What is the ratio of dimes to nickels expressed as a fraction in simplest form?

| Heather's Coins | |
|---|---|
| Dimes | 9 |
| Nickels | 6 |
| Pennies | 8 |
| Quarters | 4 |

A. $\frac{3}{2}$

B. $\frac{2}{3}$

C. $\frac{3}{5}$

D. $\frac{2}{5}$

2. A cookie recipe calls for a ratio of 4 cups of flour to 3 cups of sugar. For each cup of flour that is used, how many cups of sugar are needed?

F. $\frac{4}{3}$ cups of sugar

G. $\frac{3}{4}$ cups of sugar

H. $\frac{2}{3}$ cup of sugar

I. $\frac{3}{7}$ cup of sugar

3. Which ratio is *not* equivalent to 5 : 8?

A. 10 out of 18

B. 15 to 24

C. $\frac{20}{32}$

D. 10 : 16

4. **SHORT ANSWER** Mrs. Wilkinson can buy a 20-ounce box of cereal for $3.60 or a 28-ounce box of cereal for $4.20. Which is the better buy? Explain your reasoning.

5. The ratio table shows how much Raymond's brother earns for working different numbers of hours. How many hours would he need to work in order to earn $176?

| Hours | 5 | 8 | 15 | ▩ |
|---|---|---|---|---|
| Earnings | $40 | $64 | $120 | $176 |

F. 18 hours

G. 20 hours

H. 22 hours

I. 24 hours

6. There are 30 students in Mr. Holland's music class. If 30% of the students play in the school band, how many students in the class play in the school band?

A. 9 students

B. 12 students

C. 15 students

D. 100 students

# Course 1 Benchmark Test – First Quarter  (continued)

**7. SHORT ANSWER** Jasmine answered 19 out of 25 questions correctly on a quiz. About what percent of her answers were correct? Explain.

**8.** Which of the following shows the rational numbers in order from least to greatest?

   **F.** 25%, 0.22, $\frac{1}{5}$

   **G.** $\frac{1}{5}$, 0.22, 25%

   **H.** 0.22, $\frac{1}{5}$, 25%

   **I.** 0.22, 25%, $\frac{1}{5}$

**9.** The model below represents 66%. What decimal is equivalent to this percent?

   **A.** 0.066

   **B.** 0.66

   **C.** 6.6

   **D.** 66

**10.** How is the decimal 0.55 written as a fraction in simplest form?

   **F.** $\frac{55}{100}$

   **G.** $\frac{11}{20}$

   **H.** $\frac{11}{50}$

   **I.** $\frac{11}{55}$

**11.** Caleb's receipt for lunch is shown below. If Caleb pays with a $10 bill, how much change will he receive?

| Donna's Deli | |
| --- | --- |
| Chicken Sandwich.............. ..... | $3.79 |
| Soup.................................... ......... | $1.45 |
| Drink.................................... ....... | $1.29 |
| Tax...................................... ........... | $0.39 |

   **A.** $6.92

   **B.** $5.74

   **C.** $3.18

   **D.** $3.08

**12.** Colleen rode her bicycle 9.5 miles in 0.8 hour. What was her average speed in miles per hour?

   **F.** 11.875 miles per hour

   **G.** 10.3 miles per hour

   **H.** 8.7 miles per hour

   **I.** 7.6 miles per hour

# Course 1 Benchmark Test – First Quarter (continued)

13. There are 25 servings in a 30.2-ounce jar of peanut butter. How many ounces of peanut butter are there in 1 serving?

   **A.** 1.208 ounces

   **B.** 1.15 ounces

   **C.** 0.828 ounce

   **D.** 0.64 ounce

14. **SHORT ANSWER** Angela earns $6.25 per hour babysitting. Estimate how much she will earn if she babysits for 9 hours this weekend. Explain.

15. Which of the following is *not* a factor of 84?

   **F.** 2

   **G.** 3

   **H.** 7

   **I.** 8

16. Which of the following does *not* represent the shaded portion of the figure below?

   **A.** 25%

   **B.** $\frac{3}{4}$

   **C.** 75%

   **D.** 0.75

17. Roberta uses 5 beads for every 3 inches of string while making necklaces. Which graph best represents the ratio of necklace string to beads used?

   **F.**

   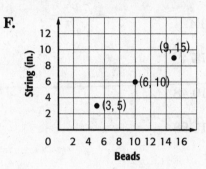

   **G.**

   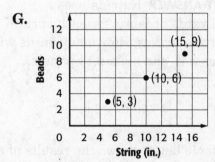

   **H.**

   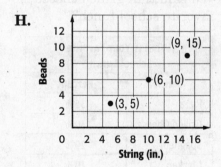

   **I.**

   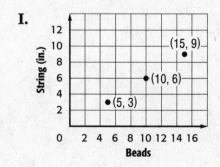

# Course 1 Benchmark Test – First Quarter  (continued)

**18.** Regina purchased 1.75 pounds of turkey breast from her local deli for $5.99 per pound. To the nearest cent, how much did she spend in all?

   **A.** $3.42

   **B.** $7.74

   **C.** $10.48

   **D.** $11.98

**19. SHORT ANSWER** Katrina uses 4.125 yards of fabric for each curtain panel she makes. How many yards will she need if she makes 14 panels?

**20.** The table below shows the results of a survey on students' favorite school lunches. What fraction of the students surveyed said that grilled cheese is their favorite school lunch?

| What Is Your Favorite School Lunch? | |
| --- | --- |
| **Lunch** | **Percent** |
| Pizza | 35% |
| Grilled Cheese | 30% |
| Spaghetti | 20% |
| Chicken | 10% |
| Soup | 5% |

   **F.** $\frac{3}{100}$

   **G.** $\frac{1}{5}$

   **H.** $\frac{1}{4}$

   **I.** $\frac{3}{10}$

**21.** Tommy's batting average this season is 0.275. This means that he had a hit in 27.5% of his at-bats. If Tommy had 11 hits so far this season, how many at-bats has he had?

   **A.** 40 at-bats

   **B.** 35 at-bats

   **C.** 26 at-bats

   **D.** 3 at-bats

**22.** In a machine, a large gear completes a revolution every minute while a small gear completes a revolution every 24 seconds. If the gears are currently aligned, how much time will pass before they are aligned again?

   **F.** 12 seconds

   **G.** 24 seconds

   **H.** 1 minute

   **I.** 2 minutes

**23.** Which of the following is the best estimate for the problem shown below?

$$151 \div 29$$

   **A.** about 4

   **B.** about 5

   **C.** about 6

   **D.** about 7

# Course 1 Benchmark Test – First Quarter *(continued)*

**24. SHORT ANSWER** One acre is equivalent to 43,560 square feet. If an acre is also equivalent to 4,840 square yards, how many square feet are equal to one square yard?

**25.** Which model represents 140%?

F.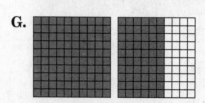

G.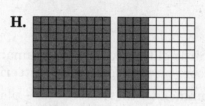

H.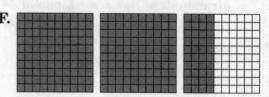

I.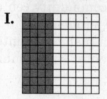

# Course 1 Benchmark Test – Second Quarter (Chapters 4–6)

1. Raul is making a scale model of an airplane that has a wingspan of 44 feet. If Raul's scale model is $\frac{1}{16}$ the size of the actual airplane, what is the wingspan of his model?

   A. 704 ft

   B. 60 ft

   C. $2\frac{3}{4}$ ft

   D. $1\frac{2}{3}$ ft

2. Two-thirds of the students in Hannah's homeroom plan to do some volunteering this summer. Of these students, $\frac{3}{5}$ plan to volunteer at the community center. What fraction of the students in Hannah's homeroom plan to volunteer at the community center this summer?

   F. $\frac{2}{3}$

   G. $\frac{3}{5}$

   H. $\frac{2}{5}$

   I. $\frac{1}{15}$

3. **SHORT ANSWER** Which point on the number line is closest to the product of the numbers graphed at points $R$ and $T$? Explain your answer.

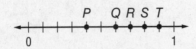

4. In which quadrant does point A lie on the coordinate plane?

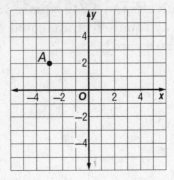

   A. I

   B. II

   C. III

   D. IV

5. Which of the following integers has the greatest absolute value?

   F. 0

   G. 7

   H. −10

   I. 1

6. The Panthers football team lost 4 yards on each of their first two plays of the game. Which of the following integers represents the progress of the team after the first two plays?

   A. −8

   B. −4

   C. 4

   D. 8

# Course 1 Benchmark Test – Second Quarter  (continued)

**7.** The table shows the record low temperatures of four different towns. Which of the following shows the record temperatures ordered from least to greatest?

| Record Low Temperatures | |
|---|---|
| **Town** | **Temperature (°F)** |
| Oakmont | −7 |
| Cherry Grove | 3 |
| Anderson Hills | 11 |
| Glentown | −2 |

**F.** 11, 3, −2, −7

**G.** −2, 3, −7, 11

**H.** −2, −7, 3, 11

**I.** −7, −2, 3, 11

**8.** Which of the following expressions correctly uses exponents to show the prime factorization of 360?

**A.** $2^4 \times 3^2 \times 5$

**B.** $2^3 \times 3^2 \times 5$

**C.** $2^4 \times 3 \times 5$

**D.** $2^3 \times 3 \times 5^2$

**9.** The expression $\frac{d}{t}$ can be used to find the average speed of an object that travels a distance $d$ in time $t$. What is a car's average speed if it travels 145 miles in 2.5 hours?

**F.** 58 miles per hour

**G.** 62 miles per hour

**H.** 65 miles per hour

**I.** 362.5 miles per hour

**10.** Which of the following expressions is equivalent to $6(5 + 3x)$?

**A** $30 + 3x$

**B** $11 + 9x$

**C** $30 + 18x$

**D** $11 + 3x$

**11.** **SHORT ANSWER** Graph and label point $W(4, -1)$ on the coordinate plane.

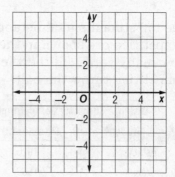

**12.** What are the coordinates of the point in Quadrant IV on the coordinate plane?

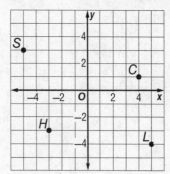

**F.** $(4, 1)$

**G.** $(1, 4)$

**H.** $(-4, 5)$

**I.** $(5, -4)$

# Course 1 Benchmark Test – Second Quarter  (continued)

**13.** Which of the following rational numbers represents a repeating decimal?

**A.** $\frac{25}{48}$

**B.** $\frac{11}{40}$

**C.** $\frac{7}{32}$

**D.** $\frac{3}{25}$

**14.** The top students in a distance throwing competition are shown in the table. How many yards did the winner of the competition throw the ball?

| Distance Throwing Competition | |
| --- | --- |
| Student | Distance (ft) |
| Ashley | 162 |
| Craig | 156 |
| Fernando | 175 |
| Robert | 166 |

**F.** 525 yards

**G.** 468 yards

**H.** $58\frac{1}{3}$ yards

**I.** 52 yards

**15. SHORT ANSWER** Define a variable and write an expression to represent the following phrase.

*seven years younger than Lisa*

**16.** Mrs. Rome has $\frac{2}{3}$ of a pan of lasagna left after dinner. She wants to divide the leftover lasagna into 4 equal servings. What fraction of the original pan does each serving represent?

**A.** $\frac{1}{12}$

**B.** $\frac{1}{6}$

**C.** $\frac{1}{4}$

**D.** $\frac{3}{8}$

**17.** Jeff is making fruit punch for the school dance. He needs $3\frac{3}{4}$ cups of pineapple juice per batch. If Jeff wants to make $4\frac{1}{2}$ batches of punch, how many cups of pineapple juice will he need?

**F.** $8\frac{1}{4}$ cups

**G.** $12\frac{3}{8}$ cups

**H.** $15\frac{1}{2}$ cups

**I.** $16\frac{7}{8}$ cups

**18.** Which of the following symbols, when placed in the blank, makes the number sentence true?

$$\frac{11}{12} \underline{\hspace{1cm}} 0.916666...$$

**A.** +

**B.** =

**C.** <

**D.** >

# Course 1 Benchmark Test – Second Quarter (continued)

**19. SHORT ANSWER** A kindergarten teacher has $22\frac{1}{2}$ cups of juice to be divided equally among her students. If each student is to receive $1\frac{1}{4}$ cups of juice, how many students are there?

**20.** A plumber has 28 feet of PVC pipe that he wants to cut into sections that are $2\frac{1}{3}$ feet long. How many sections of pipe will the plumber have in all?

**F.** $14\frac{1}{3}$ sections

**G.** $13\frac{1}{2}$ sections

**H.** 12 sections

**I.** 11 sections

**21.** Which property is represented by the equation below?

$$\frac{2}{3} \times \frac{3}{2} = 1$$

**A.** Multiplicative Inverse Property

**B.** Multiplicative Identity Property

**C.** Distributive Property

**D.** Commutative Property of Multiplication

**22.** Alexandria is evaluating the expression below.

$$3 \times 8 \div 2 + (4 - 1)^2$$

Which operation should be performed first according to the order of operations?

**F.** Multiply 3 and 8.

**G.** Divide 8 by 2.

**H.** Subtract 1 from 4.

**I.** Evaluate the power.

**23.** Which of the following coordinate planes correctly shows point $G(4, -5)$ graphed?

**A.**

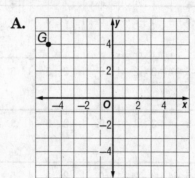

**B.**

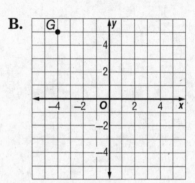

**C.**

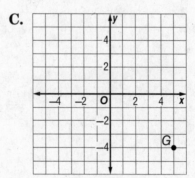

**D.**

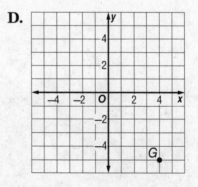

# Course 1 Benchmark Test – Second Quarter *(continued)*

**24.** Which number line shows two different integers with the same absolute value?

**F.**
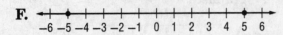
−6 −5 −4 −3 −2 −1  0  1  2  3  4  5  6

**G.**
−6 −5 −4 −3 −2 −1  0  1  2  3  4  5  6

**H.**
−6 −5 −4 −3 −2 −1  0  1  2  3  4  5  6

**I.**
−6 −5 −4 −3 −2 −1  0  1  2  3  4  5  6

**25. SHORT ANSWER** Use the Distributive Property to write a numerical expression that is equivalent to 25 + 10.

# Course 1 Benchmark Test – Third Quarter (Chapters 7–9)

**1.** The algebra mat below models the equation $x + 3 = 5$.

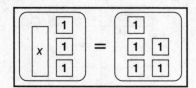

What is the solution to the equation?

**A.** −8

**B.** −2

**C.** 2

**D.** 8

**2.** The sixth graders at Lakota Middle School hope to raise *at least* $250 for a local charity. Which of the following inequalities best represents this situation?

**F.** $n \geq 250$

**G.** $n > 250$

**H.** $n < 250$

**I.** $n \leq 250$

**3.** **SHORT ANSWER** Paco's age divided by 3 is equal to 4. Let $p$ represent Paco's age and write an equation to model the situation. Then solve the equation to find Paco's age.

**4.** What is the area of the triangle shown below?

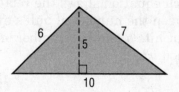

**A.** 23 square units

**B.** 25 square units

**C.** 42 square units

**D.** 50 square units

**5.** Which number line shows the solution to the inequality $x - 1 > 2$?

**F.** 
```
◄───┼──┼──┼──┼──┼──●──┼──┼──┼──┼──►
   −5 −4 −3 −2 −1  0  1  2  3  4  5
```

**G.** 
```
◄───┼──┼──┼──┼──┼──●━━┿━━┿━━┿━━►
   −5 −4 −3 −2 −1  0  1  2  3  4  5
```

**H.** 
```
◄━━━┿━━┿━━┿━━┿━━┿━━┿━━┿━━○──┼──┼──►
   −5 −4 −3 −2 −1  0  1  2  3  4  5
```

**I.** 
```
◄───┼──┼──┼──┼──┼──┼──┼──○━━┿━━┿━━►
   −5 −4 −3 −2 −1  0  1  2  3  4  5
```

**6.** Which operation should be performed to solve the equation below?

$$x - 8 = 11$$

**A.** Divide each side by 8.

**B.** Multiply each side by 8.

**C.** Add 8 to each side.

**D.** Subtract 8 from each side.

# Course 1 Benchmark Test – Third Quarter  (continued)

**7.** Grant earns 10 points for each bull's-eye that he hits in an archery competition. Which equation shows the relationship between the number of bull's-eyes Grant hits $b$ and the number of points earned $n$?

**F.** $n = \dfrac{b}{10}$

**G.** $n = 10b$

**H.** $b = 10n$

**I.** $b = \dfrac{10}{n}$

**8.** Joshua has a rectangular vegetable garden with the dimensions shown below.

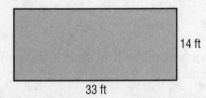

14 ft

33 ft

If he doubles the length and width of the garden, how will the area of the new garden compare to the area of the original garden?

**A.** The area of the new garden will be 4 times larger.

**B.** The area of the new garden will be 2 times larger.

**C.** The area of the new garden will be half as large.

**D.** The area of the new garden will be one fourth as large.

**9.** What value of $x$ results in a true number sentence in the equation?

$$\frac{x}{6} = 2$$

**F.** 3

**G.** 4

**H.** 8

**I.** 12

**10.** What is the missing rule in the function table?

| x | ? |
|---|---|
| 0 | 0 |
| 1 | 1 |
| 3 | 9 |
| 4 | 16 |
| 5 | 25 |

**A.** $-3x$

**B.** $-2x$

**C.** $2x$

**D.** $x^2$

**11. SHORT ANSWER** Complete the function table shown below.

| Input (x) | Output (x + 8) |
|-----------|----------------|
| 1 | |
| 4 | |
| 7 | |
| 10 | |

**12.** Which inequality is graphed on the number line?

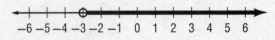

**F.** $x > -3$

**G.** $x \geq -3$

**H.** $x \leq -3$

**I.** $x < -3$

# Course 1 Benchmark Test – Third Quarter   (continued)

**13. SHORT ANSWER** What is the area of the shaded parallelogram shown below? Explain your answer.

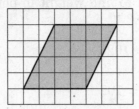

**14.** Cameron earns $6 per hour walking dogs in his neighborhood. Which of the following does *not* represent the relationship between the number of hours $x$ and Cameron's earnings $y$?

**A.** Input: $x$, Output: $6x$

**B.** $x = 6y$

**C.**

| x | 1 | 2 | 3 | 4 |
|---|---|---|----|----|
| y | 6 | 12 | 18 | 24 |

**D.**

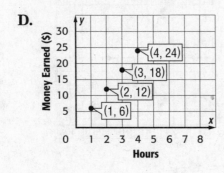

**15.** Eric spent $24 on raffle tickets. If each ticket cost $3, which equation could be used to find the number of tickets $t$ that Eric purchased?

**F.** $3t = 24$

**G.** $24t = 3$

**H.** $\dfrac{t}{3} = 24$

**I.** $\dfrac{t}{24} = 3$

**16.** What type of polygon is shown on the coordinate plane below?

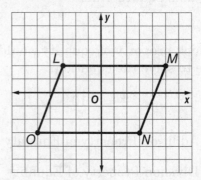

**A.** rectangle

**B.** square

**C.** rhombus

**D.** parallelogram

**17.** What is the area of the trapezoid shown below?

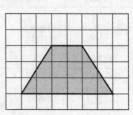

**F.** 24 square units

**G.** 15 square units

**H.** 12 square units

**I.** 9.5 square units

# Course 1 Benchmark Test – Third Quarter  *(continued)*

**18.** What is the area of the composite figure shown below?

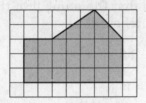

  **A.** 31 square units

  **B.** 28.5 square units

  **C.** 26 square units

  **D.** 22.5 square units

**19.** Which property of equality would you use to solve the equation?

$$\frac{w}{5} = -10$$

  **F.** Addition Property of Equality

  **G.** Division Property of Equality

  **H.** Subtraction Property of Equality

  **I.** Multiplication Property of Equality

**20.** Which rule best describes the relationship shown in the function table?

| Input | Output |
|-------|--------|
| 4 | 7 |
| 8 | 11 |
| 12 | 15 |
| 16 | 19 |
| 20 | 23 |

  **A.** subtract 3

  **B.** add 3

  **C.** divide by 2

  **D.** multiply by 2

**21.** Crystal's dog and puppy weigh 102 pounds combined. If Crystal's dog weighs 87 pounds, which equation could be used to find weight in pounds the puppy $p$?

  **F.** $\dfrac{p}{87} = 102$

  **G.** $87p = 102$

  **H.** $87 - p = 102$

  **I.** $87 + p = 102$

**22.** **SHORT ANSWER** Meredith burns 12 calories per minute while jogging. At this rate, how long will it take her to burn 180 calories? Write and solve an equation to represent this situation.

**23.** Which equation represents the function graphed on the coordinate plane?

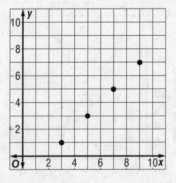

  **A** $y = x + 2$

  **B** $y = x - 2$

  **C** $x = y - 2$

  **D** $x = 2 - y$

# Course 1 Benchmark Test – Third Quarter *(continued)*

**24.** Coach Wilson hopes to have *no more than* 15 turnovers during today's basketball game. Which inequality best represents this situation?

   **F.** $t \geq 15$

   **G.** $t > 15$

   **H.** $t < 15$

   **I.** $t \leq 15$

**25. SHORT ANSWER** Draw and label a triangle that has an area of 40 square units.

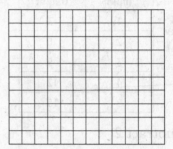

# Course 1 Benchmark Test – End of Year

**1.** Which rule best describes the relationship shown in the function table below?

| Input | Output |
|-------|--------|
| 1 | 3 |
| 2 | 6 |
| 3 | 9 |
| 4 | 12 |
| 5 | 15 |

**A.** subtract 2

**B.** add 2

**C.** divide by 3

**D.** multiply by 3

**2.** Marcus needs to earn a grade *higher than* 88 on his final quiz in order to have an A average. Which inequality best represents this situation?

**F.** $g \geq 88$

**G.** $g > 88$

**H.** $g < 88$

**I.** $g \leq 88$

**3. SHORT ANSWER** Define a variable and write an expression to represent the following phrase.

*a number increased by 5*

**4.** What is the least common multiple of 8 and 14?

**A.** 56

**B.** 28

**C.** 4

**D.** 2

**5.** What is the volume of the rectangular prism shown below?

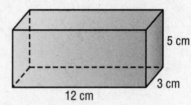

**F.** 20 cm³

**G.** 75 cm³

**H.** 180 cm³

**I.** 222 cm³

**6.** The list below shows the number of books read by students in Abram's class over the summer. What is the mode of the data?

3, 6, 12, 4, 3, 5, 4, 8, 4, 10, 4, 8, 7, 5, 7

**A.** 4 books

**B.** 5 books

**C.** 7 books

**D.** 9 books

# Course 1 Benchmark Test – End of Year (continued)

**7.** Which type of data display would be best for showing how data change over time?

 **F.** box plot

 **G.** histogram

 **H.** line graph

 **I.** line plot

**8.** There are 65 people watching a movie at a theater. If 40% of the customers purchased refreshments for the movie, how many customers purchased refreshments?

 **A.** 26 customers

 **B.** 34 customers

 **C.** 39 customers

 **D.** 163 customers

**9.** Adeline is wrapping a gift for her mother in a box with the dimensions shown.

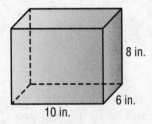

8 in.

6 in.

10 in.

What is the minimum amount of wrapping paper Adeline will need to completely cover the gift box?

 **F.** 188 square inches

 **G.** 376 square inches

 **H.** 424 square inches

 **I.** 488 square inches

**10.** The ratio table shows the number of miles Karen can drive for 1, 2, 3, and 4 gallons of gasoline. Based on the table, how far would she be able to drive on 8 gallons of gasoline?

| Gallons | 1 | 2 | 3 | 4 |
|---|---|---|---|---|
| Distance (mi) | 30 | 60 | 90 | 120 |

 **A.** 30 mi

 **B.** 150 mi

 **C.** 210 mi

 **D.** 240 mi

**11. SHORT ANSWER** Emily made 14 out of 19 shots during basketball practice. About what percent of her shots did she make? Explain your reasoning.

**12.** A muffin recipe calls for a ratio of 5 cups of flour to 2 cups of sugar. For each cup of sugar that is used, how many cups of flour are needed?

 **F.** $\frac{5}{2}$ cups of flour

 **G.** $\frac{5}{7}$ cups of flour

 **H.** $\frac{2}{5}$ cup of flour

 **I.** $\frac{2}{7}$ cup of flour

# Course 1 Benchmark Test – End of Year *(continued)*

**13. SHORT ANSWER** The line graph shows the number of members during the first few months of a photography club. Describe the data. Then predict the number of members for the sixth month.

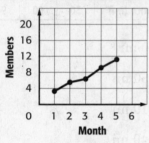

**14.** The table shows the number of points Anna scored this season. Find the mean number of points Anna scored.

| Points Scored | | | |
|---|---|---|---|
| 12 | 7 | 9 | 10 |
| 16 | 6 | 8 | 15 |
| 12 | 11 | 12 | 14 |

**A.** 9 points

**B.** 10 points

**C.** 11 points

**D.** 12 points

**15.** Which of the following integers has the least absolute value?

**F.** −3

**G.** 4

**H.** 8

**I.** −12

**16.** Albert purchased 2.4 pounds of mixed nuts for $4.79 per pound. How much did he spend in all, to the nearest cent?

**A.** $12.43

**B.** $11.50

**C.** $6.71

**D.** $1.99

**17.** Which of the following coordinate pairs corresponds to point *A*?

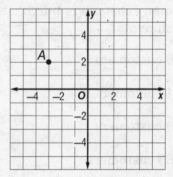

**F.** $(2, -3)$

**G.** $(3, -2)$

**H.** $(-2, 3)$

**I.** $(-3, 2)$

**18.** Which of the following symbols, when placed in the blank, makes the number sentence true?

$$\frac{20}{75} \underline{\quad} 0.\overline{26}$$

**A.** +

**B.** =

**C.** <

**D.** >

# Course 1 Benchmark Test – End of Year (continued)

**19.** What is the volume of the triangular prism?

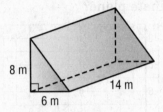

8 m
14 m
6 m

**F.** 336 cubic meters

**G.** 384 cubic meters

**H.** 672 cubic meters

**I.** 724 cubic meters

**20.** The line plot shows the quiz scores of several students.

**Quiz Scores**

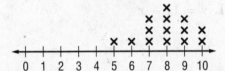

0 1 2 3 4 5 6 7 8 9 10

What is the range of the quiz scores?

**A.** 4 points

**B.** 5 points

**C.** 7 points

**D.** 8 points

**21.** Julio is evaluating the expression below.

$$6 + 2(9 - 4) - 3 \times 5$$

Which operation should be performed first according to the order of operations?

**F.** Add 6 and 2.

**G.** Multiply 2 by 9.

**H.** Subtract 4 from 9.

**I.** Multiply 3 by 5.

**22.** Which property is represented by the equation shown below?

$$6 \times 3 = 3 \times 6$$

**A.** Multiplicative Inverse Property

**B.** Multiplicative Identity Property

**C.** Associative Property of Multiplication

**D.** Commutative Property of Multiplication

**23.** The algebra mat below models the equation $x - 2 = 4$.

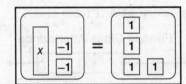

What is the solution to the equation?

**F.** 6

**G.** 2

**H.** −2

**I.** −8

**24. SHORT ANSWER** Graph the figure with the vertices $A(2, -1)$, $B(6, -1)$, and $C(6, 4)$. Then classify the figure.

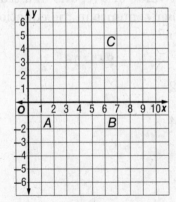

# Course 1 Benchmark Test – End of Year (continued)

**25.** Which number line shows the solution to the inequality $x + 3 \leq 1$?

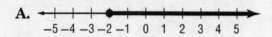 **A.**

 **B.**

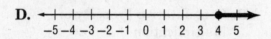 **C.**

**D.**

**26.** The box plot shows the daily attendance at a fitness class.

**Fitness Class Attendance**

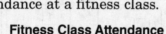
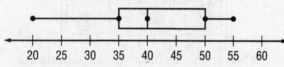

What is the median of the data?

**F.** 55

**G.** 40

**H.** 35

**I.** 20

**27.** What value of $x$ results in a true number sentence in the equation shown?

$$2x = 16$$

**A.** 32

**B.** 14

**C.** 8

**D.** 4

**28.** Which of the following equations represents the function graphed on the coordinate plane?

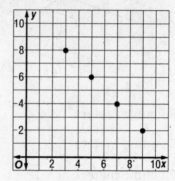

**F.** $y = x + 5$

**G.** $y = x + 1$

**H.** $y = 11 + x$

**I.** $y = 11 - x$

**29. SHORT ANSWER** The table below shows computer prices at an electronics store.

| Computer Prices ($) | | | |
|---|---|---|---|
| 950 | 620 | 545 | 810 |
| 775 | 1,120 | 905 | 775 |

Find the mean absolute deviation to the nearest cent. Explain what this value represents.

# Course 1 Benchmark Test – End of Year (continued)

**30.** The table below shows the type and number of vehicles in a parking lot.

| Types of Cars | |
|---|---|
| Minivans | 12 |
| Sedan | 28 |
| SUV | 9 |
| Trucks | 5 |

What is the ratio of sedans to minivans in simplest form?

**A.** 7 to 3

**B.** 3 to 7

**C.** 7 to 10

**D.** 10 to 3

**31.** The expression $rt$ can be used to find the distance traveled by an object that has an average speed of $r$ over time $t$. How many miles will a hot air balloon travel in 2.2 hours if it travels at an average speed of 12.5 miles per hour?

**F.** 30.1 miles

**G.** 27.5 miles

**H.** 14.7 miles

**I.** 5.7 miles

**32.** What is the area of the triangle?

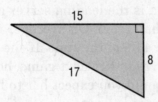

**A.** 120 square units

**B.** 75 square units

**C.** 60 square units

**D.** 40 square units

**33.** **SHORT ANSWER** The table below shows the number of canoes rented from Outdoor Adventures over the past four weekends.

| Canoe Rentals | | | |
|---|---|---|---|
| 21 | 32 | 17 | 24 |
| 15 | 30 | 28 | 26 |

Find the range, median, first quartile, third quartile, and interquartile range of the data.

**34.** A carpenter makes 4 table legs for each table that he builds. Which equation represents the relationship between the number of tables built $t$ and the number of legs made $l$?

**F.** $l = 4t$

**G.** $t = 4l$

**H.** $l = t + 4$

**I.** $t = l + 4$

**35.** Which of the following ratios is equivalent to $\frac{5}{8}$?

**A.** $16 : 10$

**B.** 5 to 13

**C.** $\frac{25}{44}$

**D.** 15 out of 24

# Course 1 Benchmark Test – End of Year    (continued)

**36.** Kylie surveyed several classmates about the number of states they have visited. The results are shown in the histogram.

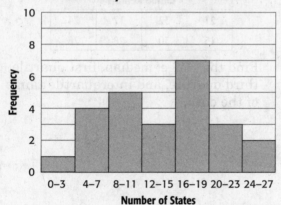

**How Many States Have You Visited?**

How many of Kylie's classmates have visited more than 15 states?

**F.** 3 students

**G.** 8 students

**H.** 12 students

**I.** 15 students

**37.** What is the surface area of the triangular prism?

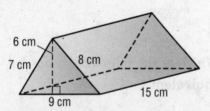

**A.** 468 square centimeters

**B.** 414 square centimeters

**C.** 405 square centimeters

**D.** 378 square centimeters

**38.** Which of the following represents the decimal 0.32 written as a fraction in simplest form?

**F.** $\frac{32}{100}$

**G.** $\frac{16}{50}$

**H.** $\frac{17}{50}$

**I.** $\frac{8}{25}$

**39. SHORT ANSWER** Jeremy can purchase a 1.2-pound package of ground beef for $4.55 or a 1.6-pound package for $6.30. Which is the better buy? Explain your reasoning.

**40.** Pamela is the leading server on her volleyball team. On average, she serves an ace 44% of the time. If she attempts 25 serves in her next game, how many aces would you expect her to have?

**A.** 57 aces

**B.** 19 aces

**C.** 11 aces

**D.** 8 aces

# Course 1 Benchmark Test – End of Year  (continued)

**41.** What percent is represented by the model?

    **F.** 175%

    **G.** 125%

    **H.** 75%

    **I.** 25%

**42.** Which of the following best describes the center of a data set if there are outliers in the data but no big gaps in the middle of the data?

    **A.** mean

    **B.** median

    **C.** mode

    **D.** range

**43. SHORT ANSWER** Complete the function table.

| Input (x) | Output (3x – 1) |
|-----------|------------------|
| 1 | |
| 2 | |
| 3 | |
| 4 | |
| 5 | |

**44.** What is the surface area of a square pyramid with base side lengths of 10 inches and a slant height of 14 inches?

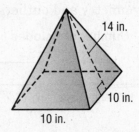

    **F.** 220 in$^2$

    **G.** 280 in$^2$

    **H.** 380 in$^2$

    **I.** 660 in$^2$

**45.** Which of the following properties would you use to solve the equation?

$$r + 4 = 11$$

    **A.** Addition Property of Equality

    **B.** Division Property of Equality

    **C.** Multiplication Property of Equality

    **D.** Subtraction Property of Equality

**46.** Which of the following inequalities is graphed on the number line?

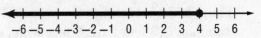

    **F.** $x > 4$

    **G.** $x \geq 4$

    **H.** $x \leq 4$

    **I.** $x < 4$

# Course 1 Benchmark Test – End of Year   (continued)

**47. SHORT ANSWER** The box plot below shows the number of Calories in different lunches at a restaurant. Describe the shape of the distribution using symmetry and outliers.

**Number of Calories**

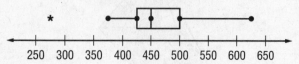

**48.** What is the area of trapezoid *QRST*?

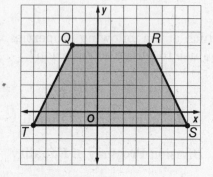

   **A.** 54 square units
   **B.** 68 square units
   **C.** 76 square units
   **D.** 108 square units

**49.** Mr. Addison is building a sandbox shaped like a rectangular prism. The sandbox is 8 feet long, 6 feet wide, and 1.5 feet deep. How many cubic feet of sand will the sandbox hold?

   **F.** 15.5 cubic feet
   **G.** 72 cubic feet
   **H.** 105 cubic feet
   **I.** 138 cubic feet

**50.** The Pirates football team has played 75% of its games so far this season. If the team has played 9 games, how many games are there in the season?

   **A.** 7 games

   **B.** 11 games

   **C.** 12 games

   **D.** 15 games

**51.** Which of the following expressions is equivalent to $3(4x + 1)$?

   **F.** $7x + 4$

   **G.** $x + 4$

   **H.** $12x + 1$

   **I.** $12x + 3$

**52.** What is the missing rule in the function table?

| x | ? |
|---|---|
| 2 | 7 |
| 3 | 8 |
| 6 | 11 |
| 9 | 14 |
| 12 | 17 |

   **A.** $\dfrac{x}{-4}$

   **B.** $x + 5$

   **C.** $-4x$

   **D.** $x - 3$

# Course 1 Benchmark Test – End of Year (continued)

**53.** Which of the following expressions correctly uses exponents to show the prime factorization of 168?

   **F.** $2^4 \times 3 \times 7$

   **G.** $2^3 \times 3^2 \times 7$

   **H.** $2^4 \times 3^2 \times 7$

   **I.** $2^3 \times 3 \times 7$

**54. SHORT ANSWER** Which measure of center would you use to describe the center of the data shown on the line plot? Explain your reasoning.

**Number of Pets**

**55.** A pancake recipe calls for $\frac{1}{3}$ cup of mix for 4 pancakes. If Beth needs to make 60 pancakes, how many cups of pancake mix will she need?

   **A.** 5 cups

   **B.** $4\frac{2}{3}$ cups

   **C.** $3\frac{1}{3}$ cups

   **D.** $\frac{1}{5}$ cup

# Chapter 1 Answer Key

## Are You Ready?—Review
### Page 1

1. _____ 20
2. _____ 12
3. _____ 86
4. _____ 28
5. _____ 26
6. _____ 18
7. _____ 12
8. _____ 17

## Are You Ready?—Practice
### Page 2

1. _____ 19
2. _____ 13
3. _____ 11
4. _____ 6
5. _____ $26
6. _____ 99 miles
7. _____ $\dfrac{1}{3}$
8. _____ $\dfrac{7}{9}$
9. _____ $\dfrac{2}{9}$
10. _____ $\dfrac{6}{7}$
11. _____ $\dfrac{11}{28}$
12. _____ $\dfrac{1}{6}$

# Chapter 1 Answer Key

1. **FIELD TRIP** A group of 92 students went on a field trip to a nature preserve. When they arrived, they were separated into 4 groups. If each group had the same number of students, how many students were in each group? **23 students**

2. **CARS** The 91 new cars in a dealership are arranged into 7 rows. If each row has the same number of cars, how many cars are in each row? **13 cars**

3. **EARNINGS** Ming earned $336 from babysitting over the past 6 weeks. She earned the same amount of money each week. How much money did she earn each week? **$56**

4. **GASOLINE** The Harnett family used 18 of the 26 gallons of gasoline they purchased. What fraction of the gasoline, in simplest form, did they use?
$\dfrac{9}{13}$

5. **CAR WASH** The student council waxed 27 of the 63 cars they washed. What fraction of the cars, in simplest form, did they wax?
$\dfrac{3}{7}$

6. **BARBEQUE** Mr. Salcido bought 24 hot dogs and 36 hamburgers for a barbeque. What fraction, in simplest form, of his food items are hot dogs?
$\dfrac{2}{5}$

# Chapter 1 Answer Key

**Diagnostic Test**
**Page 4**

1. _____ **29** _____

2. _____ **19** _____

3. _____ **21** _____

4. _____ **31** _____

5. _____ **$76** _____

6. _____ $\dfrac{1}{5}$ _____

7. _____ $\dfrac{1}{19}$ _____

8. _____ $\dfrac{7}{9}$ _____

9. _____ $\dfrac{1}{5}$ _____

10. _____ $\dfrac{3}{7}$ _____

**Pretest**
**Page 5**

1. _____ $\dfrac{3}{5}$ _____

2. _____ $\dfrac{4}{5}$ _____

3. _____ **5 to 8** _____

4. _____ **1 to 4** _____

5. _____ **45 words per minute** _____

6. _____ **$24** _____

7. _____ **16 cups** _____

Answers

# Chapter 1 Answer Key

**Chapter Quiz**
Page 6

1. _____13_____

2. _____21_____

3. $\frac{3}{4}$; For every 3 cats, there are 4 dogs.

4. $\frac{4}{5}$; For every 4 girls, there are 5 boys in the band.

5. $\frac{\$4}{1 \text{ pound}}$

6. $\frac{40 \text{ mi}}{1 \text{ h}}$

7. _____\$1.50_____

8.
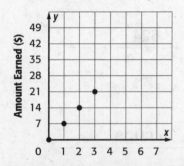
The graph shows that Dawson earned \$7 for every day he watched his neighbor's cat.

9. _____\$127_____

**Vocabulary Test**
Page 7

1. Equivalent ratios

2. unit rate

3. scaling

4. ratio table

5. least common multiple (LCM)

6. x-coordinate

7. greatest common factor

8. y-coordinate

9. Sample answer: A comparison of two quantities by division.

10. Sample answer: A ratio that compares two quantities with different kinds of units.

# Chapter 1 Answer Key

**Student Recording Sheet, Page 10**

*Use this recording sheet with the Standardized Test Practice.*

**Fill in the correct answer. For gridded-response questions, write your answers in the boxes on the answer grid and fill in the bubbles to match your answers.**

1. Ⓐ ● Ⓒ Ⓓ

2. Ⓕ ● Ⓗ Ⓘ

3.
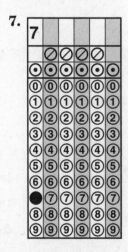

250/11

4. ● Ⓑ Ⓒ Ⓓ

5. Ⓕ ● Ⓗ Ⓘ

6. Ⓐ Ⓑ Ⓒ ●

7.

7

8. Ⓕ Ⓖ ● Ⓘ

**The ratio of boys to girls before the new students enrolled was 4:5. After the new students, the ratio changed to 15:16. The ratio did not stay the same.**

9. _____

10. Ⓐ Ⓑ ● Ⓓ

11.

2

12. $\frac{11}{16}$ mi

13. Ⓕ ● Ⓗ Ⓘ

## Extended Response

Record your answers for Exercise 14 on the back of this paper.

Part A   $\frac{45}{90} = \frac{120}{\blacksquare}$

Part B   4 h

Part C   6 h; Since $\frac{30}{90} = \frac{120}{360}$, the class would take 360 minutes or 6 hours to collect 120 books.

Answers

# Chapter 1 Answer Key

**Extended-Response Test, Page 11**
**Sample Answers**

*In addition to the scoring rubric, the following sample answers may be used as guidance in evaluating extended response assessment items.*

**1. a.** A ratio is the comparison of two numbers by division.

**b.** 3 out of 4, 3:4, 3 to 4, $\frac{3}{4}$

**c.** A rate is a ratio that compares two different units of measure. A unit rate is a rate that is in the per unit form. A unit rate is written with a denominator of 1. An example of a unit rate is 50 miles in 1 hour or 50 miles per hour. An example of rate that is not a unit rate is 50 miles in 2 hours.

**d.** Equivalent ratios express the same relationship between two quantities.

**e.** Sample answer: If two out of five students in Mrs. Junkin's class have a dog, predict how many of the 200 students in the school have a dog.

**f.** $\dfrac{2}{5} = \dfrac{\blacksquare}{200}$

$\dfrac{2}{5} = \dfrac{80}{200}$  ⤻ ×40   Since 5 × 40 = 200, multiply the numerator and denominator by 40.

So, about 80 students out of the 200 students can be expected to have a dog.

**2. a.** Sample answer: The exchange rate is such that 1 euro will cost $2.00.

**b.** $4.00; $10.00; $20.00

**c.** 15; Sample answer: Divide 30 by 2.

**d.** (1, 2), (2, 4), (5, 10), (10, 20)

# Chapter 1 Answer Key

**Test, Form 1A**
**Page 13**

**Test, Form 1A** *(continued)*
**Page 14**

1. _____ **B** _____

2. _____ **H** _____

3. _____ **D** _____

4. _____ **G** _____

5. _____ **A** _____

6. _____ **G** _____

7. _____ **B** _____

8. _____ **G** _____

9. _____ **B** _____

10. _____ **F** _____

11. _____ **A** _____

12. _____ **F** _____

**Answers**

# Chapter 1 Answer Key

1. ____C____

2. ____G____

3. ____D____

4. ____H____

5. ____A____

6. ____I____

7. ____D____

8. ____G____

9. ____B____

10. ____H____

11. ____C____

12. ____I____

# Chapter 1 Answer Key

1. ___D___

2. ___I___

3. ___B___

4. ___F___

5. ___C___

6. ___G___

7. ___B___

8. ___H___

9. ___A___

10. __375 dishes__

11. __138 students__

12. __6 blankets__

13. ___540 s___

14. yes; $\dfrac{24}{4} = \dfrac{12}{2}$

15. (0, 0), (1, 3), (2, 6), (3, 9)

Answers

# Chapter 1 Answer Key

1. ____C____

2. ____F____

3. ____C____

4. ____H____

5. ____A____

6. ____G____

7. ____A____

8. ____I____

9. ____B____

10. __228 students__

11. __101 Calories__

12. __18 matches__

13. __54 mediums__

14. no; $\dfrac{\dfrac{\$1}{2\text{ days}}}{\ } \neq \dfrac{\$3}{4\text{ days}}$

15. __(0,0), (1,4), (2,8), (3,12)__

# Chapter 1 Answer Key

1. __in 12 weeks__

2. __$\frac{7}{11}$; For every 7 houses, there are 11 businesses.__

3. __$\frac{28}{123}$__

4. __$\frac{7}{18}$__

5. __$\frac{2.7 \text{ heartbeats}}{1 \text{ s}}$__

6. __$\frac{18 \text{ students}}{1 \text{ instructor}}$__

7. __$\frac{35 \text{ meters}}{1 \text{ s}}$__

8. __48 limes__

9. __28 countries__

10. __$150__

11. __Yes; the unit rates are the same, $\frac{\$7}{1 \text{ car}}$.__

12. __16 people__

13. __104 revolutions__

14. __24 costumes__

15. __(0, 0), (1, 45), (2, 90), (3, 135)__

**Answers**

# Chapter 1 Answer Key

1. in 6 weeks

2. $\frac{1}{15}$; For every empty seat there are 15 filled seats.

3. $\frac{37}{103}$

4. $\frac{5}{16}$

5. $\frac{54.5 \text{ km}}{1 \text{ h}}$

6. $\frac{\$5.50}{1 \text{ month}}$

7. $\frac{8 \text{ hot dogs}}{1 \text{ minute}}$

8. 84 mi

9. 54 days

10. $97.50

11. No; the unit rates are not the same, $\frac{\$4}{1h}$ and $\frac{\$4.50}{1h}$

12. 45 people

13. $13.50

14. 15 in.

15. (0, 0), (1, 60), (2, 120), (3, 180)

# Chapter 2 Answer Key

**Are You Ready?—Review**
Page 25

1. _____ 2
2. _____ 6
3. _____ 8
4. _____ 14
5. _____ 4
6. _____ 3
7. _____ 12
8. _____ 5
9. _____ 21
10. _____ 3

**Are You Ready?—Practice**
Page 26

1. _____ 12
2. _____ 6
3. _____ 14
4. _____ 3
5. _____ 18
6. _____ 4

7. _____ $22
8. _____ 33
9. _____ 45
10. _____ 6
11. _____ 63
12. _____ 96, 120, 144

Answers

# Chapter 2 Answer Key

1. **GARDENING** Raven wants to plant 35 marigold plants, 20 zinnia plants, and 15 daylily plants in her flower garden. If she puts the same number of plants in each row and if each row has only one type of plant, what is the greatest number of plants she can put in one row? Explain. **5 plants; 5 is the GCF of 35, 20, and 15**

2. **BAND** The high school jazz band rehearses sometimes with exactly 4 people in every row on stage. Sometimes it rehearses with exactly 5 people in every row. What is the least number of people that can be in the band? **20 people**

3. **FARMING** Omar is planting a small plot of corn for his family. He has enough seeds to plant 9 or 12 plants in each row. What is the least number of seeds Omar could have? **36 seeds**

4. **FIELD TRIP** The sixth-grade teachers collected money from students during three days for theater tickets to a play. The ticket cost a whole number of dollars. If every student paid the same amount, what is the most the tickets could cost per student? **$4**

| Monday | $64 |
|---|---|
| Tuesday | $36 |
| Wednesday | $52 |

5. **GIFTS** Yara is making gift baskets for her neighbors. She has 12 chocolate chip cookies, 9 walnut cookies, and 15 sugar cookies to put in the baskets. Each basket must have the same number of cookies in it. Without mixing cookies, what is the greatest number of cookies Yara can put in each basket? Explain. **3 cookies; 3 is the GCF of 12, 9, and 15.**

6. **CLASS PHOTO** The sixth-grade class wants to make a class photo. The students can stand in rows with 7 students in each row, or they can stand with 8 in each row. What is the least number of students that can be in the sixth-grade class? **56 students**

# Chapter 2 Answer Key

**Diagnostic Test**
**Page 28**

1. _____ 7 _____

2. _____ 5 _____

3. _____ 17 _____

4. _____ 6 _____

5. _____ 11 _____

6. _____ 9 _____

7. _____ 25 m _____

8. _____ 8 _____

9. _____ 42 _____

10. _____ 9 _____

11. _____ 56 _____

12. _____ 24 _____

13. _____ 78 _____

14. _____ 144, 180, 216 _____

**Pretest**
**Page 29**

1. _____ $\dfrac{3}{4}$ _____

2. _____ $\dfrac{2}{5}$ _____

3. _____ 0.8 _____

4. _____ 0.04 _____

5. _____ $\dfrac{8}{25}$ _____

6. _____ $\dfrac{12}{25}$ _____

7. _____ 35% _____

8. _____ 20% _____

9. _____ 0.12 _____

10. _____ 0.64 _____

11. _____ 85% _____

12. _____ 73% _____

13. _____ > _____

14. _____ < _____

15. _____ 2.28 _____

16. _____ 243.8 _____

**Answers**

# Chapter 2 Answer Key

**Chapter Quiz**
**Page 30**

1. _____ 0.875 _____

2. _ 0.467 (rounded) _

3. _____ 4.12 _____

4. _____ $\frac{9}{20}$ _____

5. _____ $\frac{1}{8}$ _____

6. _____ $6\frac{1}{25}$ _____

7. _____ $98\frac{3}{5}$ _____

8. _____ 0.21 _____

9. _____ 0.004 _____

10. _____ 35% _____

11. _____ 81.2% _____

12. _____ $\frac{1}{8}$ _____

13. _____ $\frac{12}{25}$ _____

14. _____ 35% _____

15. _____ 60% _____

16. _____ 3; 3 _____

17. _____ 0.005; $\frac{1}{200}$ _____

18. _____ 750% _____

19. _____ 6.8% _____

20. Sample answer: 90 ÷ 30, or 3 min

**Vocabulary Test**
**Page 31**

1. _ least common denominator (LCD) _

2. _ rational number _

3. _ percent _

4. _ proportion _

5. _ percent _

6. _ percent proportion _

7. Sample answer: The least number that is a multiple of the denominators.

8. Sample answer: A number that I can write as a fraction.

9. Sample answer: A ratio that compares a number to 100.

**Course 1 • Chapter 2** Fractions, Decimals, and Percents

# Chapter 2 Answer Key

## Student Recording Sheet, Page 34

*Use this recording sheet with the Standardized Test Practice pages.*

**Fill in the correct answer. For gridded-response questions, write your answers in the boxes on the answer grid and fill in the bubbles to match your answers.**

1. (A) (B) ● (D)

2. (F) (G) (H) ●

3. ● (B) (C) (D)

4.
| 7 | 2 | | | |

5. (F) ● (H) (I)

6. (A) (B) ● (D)

7.
| 1 | 6 | | | |

8. (F) ● (H) (I)

9. (A) (B) ● (D)

10. (F) (G) ● (I)

11. (A) (B) ● (D)

12. _____ $360

## Extended Response

Record your answers for Exercise 13 on the back of this paper.

**Part A** Model A

**Part B** Model B

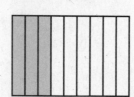

**Part C** Model B; it has $\frac{1}{3}$ shaded. $0.25 = \frac{1}{4}$ and $\frac{1}{4} < \frac{1}{3}$.

**Course 1 • Chapter 2** Fractions, Decimals, and Percents

**A17**

**Answers**

# Chapter 2 Answer Key

*In addition to the scoring rubric, the following sample answers may be used as guidance in evaluating extended response assessment items.*

1. **a.** To write a percent as a fraction, write it with a denominator of 100 and simplify. To write a percent as a decimal, rewrite the percent as a fraction with a denominator of 100 and write the fraction as a decimal.

   **b.** 30% is about $\frac{1}{3}$, $44 is about 45, $\frac{1}{3}$ of 45 is 15. The amount taken off the jacket is about $15.

2. **a.** Nov: $\frac{3}{8} = 0.375 = 37.5\%$; Dec: $\frac{1}{8} = 0.125 = 12.5\%$; Jan: $\frac{1}{2} = 0.5 = 50\%$

   **b.** Nov: 40% = 0.4; Dec: 32% = 0.32; 0.4 > 0.375; 0.32 > 0.125

3. To write a fraction as a percent, write equivalent ratios with the fraction equal to $\frac{n}{100}$. Then solve for $n$ and write as $n\%$.

# Chapter 2 Answer Key

**Test, Form 1A**
**Page 37**

1. _____ B _____

2. _____ H _____

3. _____ A _____

4. _____ G _____

5. _____ A _____

6. _____ H _____

7. _____ A _____

8. _____ H _____

9. _____ D _____

10. _____ H _____

**Test, Form 1A** *(continued)*
**Page 38**

11. _____ A _____

12. _____ I _____

13. _____ C _____

14. _____ G _____

15. _____ B _____

16. _____ I _____

17. _____ B _____

18. _____ H _____

19. _____ B _____

20. _____ H _____

**Answers**

# Chapter 2 Answer Key

**Test, Form 1B**
**Page 39**

1. B
2. F
3. D
4. F
5. C
6. F
7. C
8. F
9. A
10. G

**Test, Form 1B** *(continued)*
**Page 40**

11. D
12. I
13. D
14. H
15. C
16. H
17. D
18. G
19. C
20. F

# Chapter 2 Answer Key

1. ___A___

2. ___G___

3. ___A___

4. ___G___

5. ___B___

6. ___G___

7. ___D___

8. ___H___

9. ___B___

10. ___I___

11. ___C___

12. ___H___

13. ___D___

14. ___F___

15. ___A___

16. ___G___

17. ___28___

18. ___150___

19. ___74%___

20. ___12%___

21. $1.85;\ 1\frac{17}{20}$

22. $0.0035;\ \frac{7}{2,000}$

23. no; He made $\frac{18}{40}$ or 45% of the shots he took.

24. ___$7.20___

Answers

# Chapter 2 Answer Key

1. _____**A**_____

2. _____**H**_____

3. _____**B**_____

4. _____**I**_____

5. _____**D**_____

6. _____**F**_____

7. _____**B**_____

8. _____**G**_____

9. _____**B**_____

10. _____**F**_____

11. _____**D**_____

12. _____**H**_____

13. _____**B**_____

14. _____**F**_____

15. _____**D**_____

16. _____**H**_____

17. _____**26.35**_____

18. _____**187.2**_____

19. _____**94%**_____

20. _____**16%**_____

21. _____**1.65; $1\frac{13}{20}$**_____

22. _____**0.0065; $\frac{13}{2,000}$**_____

23. _____**yes; $\frac{7}{20}$ = 0.35 or 35%.**_____

24. _____**$7.70**_____

# Chapter 2 Answer Key

**Test, Form 3A**
**Page 45**

1. ___378___

2. No; $\frac{21}{25}$ = ___84%___

3. ___$2\frac{3}{10}$___

4. ___0.22___

5. ___yes; Her total cost is $4.82.___

6. ___$30.60___

7. ___$\frac{11}{200}$___

8. ___$\frac{7}{8}$___

9. ___35%___

10. ___70%___

11. ___2.2; $2\frac{1}{5}$___

12. ___0.0008; $\frac{1}{1,250}$___

13. ___120.36___

14. ___78.2___

15. ___100___

16. ___20___

**Test, Form 3A** *(continued)*
**Page 46**

17. ___see into the future___

18. ___read minds___

19. ___see at night___

20. ___no; Her total cost is $10.32.___

21. ___27.5%___

22. ___8%___

23. ___0.55___

24. ___0.2___

25. ___8.375___

26. ___12 times___

27. ___$8.40___

28. ___Bears, Tigers, Mustangs___

Answers

# Chapter 2 Answer Key

1. ___114___

2. yes; $\frac{9}{30} =$ 0.3 or 30%

3. $1\frac{1}{4}$

4. 0.48

5. yes; His total cost is $7.95.

6. $103.20

7. $\frac{3}{40}$

8. $\frac{5}{8}$

9. 24%

10. 25%

11. 3.45; $3\frac{9}{20}$

12. 0.0006; $\frac{3}{5,000}$

13. 124.8

14. 39

15. 85

16. 75

17. see into the future

18. read minds

19. become invisible

20. no; His total cost is $16.40.

21. 32.5%

22. 2%

23. 0.28

24. 0.625

25. 8.9

26. 9 times

27. $9

28. Mustangs, Bears, Tigers

A24

# Chapter 3 Answer Key

**Are You Ready?—Review**
**Page 49**

1. ___8___

2. ___9___

3. ___4___

4. ___9___

5. ___5___

6. ___2___

7. ___10___

8. ___11___

9. ___9___

10. ___12___

**Are You Ready?—Practice**
**Page 50**

1. ___1,426___

2. ___434___

3. ___420___

4. ___680___

5. ___1,720___

6. ___429___

7. ___$156___

8. ___$195___

9. ___936 ft___

10. ___77___

11. ___61___

12. ___161___

13. ___200___

14. ___14 h___

**Answers**

# Chapter 3 Answer Key

1. **CLOTHES** Ellen buys 6 new shirts for $22 each. How much does she spend on shirts? **$132**

2. **SHOPPING** Fruit is sold by the bushel at the farmer's market. Andrew buys 12 bushels of peaches to make preserves for the school bake sale. How much does he spend on peaches? **$156**

   | Apricots | $15 per bushel |
   |----------|----------------|
   | Peaches  | $13 per bushel |
   | Pears    | $12 per bushel |

3. **ZOO** The bears at the zoo eat 875 pounds of food each week. How much do they eat per day? **125 lb**

4. **DISTANCE** Mrs. Mendez drives 34 miles each day to take her children to school and run errands. How many miles did she drive in 13 days? **442 mi**

5. **RUNNING** The school track team ran 96 miles in 12 days. How many miles did they run per day? **8 mi**

6. **GRASS** Ernie mows grass to make money. He made $324 for mowing 4 lawns last week. If he made the same amount on each lawn, how much did he get paid for each? **$81**

# Chapter 3 Answer Key

**Diagnostic Test**
**Page 52**

1. _____ **750** _____

2. _____ **533** _____

3. _____ **544** _____

4. _____ **616** _____

5. _____ **3,720** _____

6. _____ **2,211** _____

7. _____ **$136** _____

8. _____ **$132** _____

9. _____ **1,625 ft²** _____

10. _____ **56** _____

11. _____ **27** _____

12. _____ **143** _____

13. _____ **99** _____

14. _____ **65 mph** _____

**Pretest**
**Page 53**

1. _____ **16** _____

2. _____ **300** _____

3. _____ **44.8** _____

4. _____ **63.27** _____

5. _____ **0.1884** _____

6. _____ **10.2** _____

7. _____ **7** _____

8. _____ **3** _____

9. _____ **2.7** _____

10. _____ **0.7** _____

11. _____ **17** _____

12. _____ **0.8** _____

13. _____ **66,500** _____

14. _____ **0.0432** _____

15. _____ **0.02374** _____

16. _____ **9** _____

**Answers**

# Chapter 3 Answer Key

1.  $49 \times 10$ or 490

2.  $70 \times 6$ or 420

3.  97.08

4.  4.745

5.  60 mi

6.  7.2

7.  10

8.  15.5

9.  25.2

10.  $1.75

11.  7.248

12.  0.1625

13.  11.55 in$^2$

1.  compatible numbers

2.  dividend

3.  quotient

4.  product

5.  quotient

6.  divisor

7.  product

8.  dividend

9.  Sample answer: Compatible numbers are numbers that I can divide easily.

10.  Sample answer: To annex a zero means to put an extra zero in front of a number or at the end of it without changing the number's value.

# Chapter 3 Answer Key

**Student Recording Sheet, Page 58**

*Use this recording sheet with the Standardized Test Practice pages.*

**Fill in the correct answer. For gridded-response questions, write your answers in the boxes on the answer grid and fill in the bubbles to match your answers.**

1. Ⓐ Ⓑ ● Ⓓ

2. Ⓕ ● Ⓗ Ⓘ

3. ___$1.58___

4.
| 1 | 2 | | | |

5.
| 2 | 0 | · | 4 | |

6. ● Ⓑ Ⓒ Ⓓ

7. ● Ⓖ Ⓗ Ⓘ

8.
| 1 | · | 6 | 8 | |

9. Ⓐ ● Ⓒ Ⓓ

10. $5.7 \times 0.59 = 3.363$ or $3.36$ and $2.8 \times 1.99 = 5.572$ or $5.57$; So she paid $8.93 for the bananas and apples.

11. Ⓕ Ⓖ ● Ⓘ

12. Ⓐ Ⓑ Ⓒ ●

13. Ⓕ Ⓖ ● Ⓘ

14. **Part A**  plot A: base: 20 ft, perimeter: 60.4 ft
plot B: area: 204.48 sq ft, height: 16 ft
plot C: base: 14.3 ft, perimeter: 57.2 ft

**Part B**  Plot C

**Part C**  Plot C

## Extended Response

**Record your answers for Exercise 14 on the back of this paper.**

# Chapter 3 Answer Key

## Extended-Response Test, Page 59
### Sample Answers

*In addition to the scoring rubric, the following sample answers may be used as guidance in evaluating extended response assessment items.*

**1. a.** Sample answer: Round 5.25 to 5 and 4.8 to 5. Then the approximate area of the conservation area is $5 \times 5$ or 25 square miles.

   **b.** Sample answer: Multiply as with whole numbers. Then count the number of decimal places in each factor and find their sum. The product should have the same number of decimal places as their sum. The area of the conservation area is 25.2 square miles.

   **c.** Sample answer: To divide by a decimal, change the divisor to a whole number. Do this by multiplying the divisor and dividend by the same power of ten. Then divide as with whole numbers. The length of the lot is 1.75 miles.

**2. a.** Using the estimation method, round 11.25 to 11. $11 \times 6$ is 66. Find $6 \times 11.25$ as though multiplying whole numbers. The result is 6,750. Since the estimate is 66, the decimal point goes after the 7. The answer is 67.5.

or

Using the counting decimal places method, multiply 6 and 11.25 as though they were whole numbers. The result is 6,750. Count the number of places to the right of the decimal point for each number, which is 2. Then count the same number of decimal places from right to left in the product. Place the decimal point here. The answer is 67.5.

   **b.** Divide $18.57 by $2.

```
      9.285
  2)18.570
   −18
     05
     −4
     17
    −16
     10
    −10
      0
```

They can buy only 9 packs, since they cannot buy a partial pack.

**3. a.** Since the decimal point was moved 4 places to the right, the missing factor is 10,000.

   **b.** When dividing, it is important to keep digits with the correct place value. Therefore, you annex zeros in the quotient. You can annex zeros in the dividend to continue dividing after the decimal point.

# Chapter 3 Answer Key

**Test, Form 1A**
**Page 61**

1. ___C___

2. ___G___

3. ___B___

4. ___H___

5. ___B___

6. ___I___

7. ___B___

8. ___F___

9. ___C___

10. ___H___

11. ___D___

**Test, Form 1A**  *(continued)*
**Page 62**

12. ___H___

13. ___D___

14. ___G___

15. ___C___

16. ___F___

17. ___B___

18. ___H___

19. ___C___

20. ___I___

**Answers**

# Chapter 3 Answer Key

**Test, Form 1B**
**Page 63**

1. _____ **B** _____

2. _____ **G** _____

3. _____ **C** _____

4. _____ **G** _____

5. _____ **A** _____

6. _____ **I** _____

7. _____ **D** _____

8. _____ **G** _____

9. _____ **D** _____

10. _____ **G** _____

11. _____ **B** _____

**Test, Form 1B** *(continued)*
**Page 64**

12. _____ **I** _____

13. _____ **B** _____

14. _____ **H** _____

15. _____ **D** _____

16. _____ **F** _____

17. _____ **B** _____

18. _____ **F** _____

19. _____ **B** _____

20. _____ **G** _____

# Chapter 3 Answer Key

**Test, Form 2A**
Page 65

1. _____**A**_____

2. _____**H**_____

3. _____**C**_____

4. _____**I**_____

5. _____**B**_____

6. _____**G**_____

7. _____**D**_____

8. _____**H**_____

9. _____**C**_____

10. _____**G**_____

**Test, Form 2A**  *(continued)*
Page 66

11. _____**C**_____

12. _____**G**_____

13. _____**A**_____

14. _____**$1.28**_____

15. _____**13.3**_____

16. _____**4**_____

17. _____**31.82 m²**_____

18. _____**44.1 in²**_____

19. _____**$85.15**_____

20. **14 buses; 13 buses will hold only 624 students.**

21. **14.6 ft; 14.6 × 18 = 262.8**

22. **10 pages; 9 pages will hold only 378 stickers.**

**Answers**

# Chapter 3 Answer Key

1. ___B___

2. ___H___

3. ___B___

4. ___H___

5. ___B___

6. ___G___

7. ___C___

8. ___G___

9. ___B___

10. ___I___

11. ___B___

12. ___H___

13. ___C___

14. ___$1.48___

15. ___17.2___

16. ___3___

17. ___38.95 m²___

18. ___31.2 in²___

19. ___$95.58___

20. 17 buses; 16 buses will hold only 736 students.

21. 12.6 ft; 12.6 × 16 = 201.6

22. 13 pages; 12 pages will hold only 456 stickers.

# Chapter 3 Answer Key

1. ___6 × 9 or 54___

2. ___12 × 3 or 36___

3. ___24 ÷ 8 or 3___

4. ___63 ÷ 9 or 7___

5. ___250 mi;___
   ___25 × 10 = 250___

6. ___6 bracelets;___
   ___42 ÷ 7 = 6___

7. ___14.4 mi___

8. ___$63.09___

9. ___1.35 m²___

10. ___9.432 s___

11. ___$4.39___

12. ___2.1___

13. ___0.004535___

14. ___0.602___

15. ___940___

16. ___15.5___

17. ___$695___

18. ___$11___

19. ___$1.40___

20. ___31___

21. ___4.05___

22. ___8.2___

23. ___31___

24. ___5.05___

25. ___$14.99___

26. ___3.7 oz___

27. ___0.225 mi___

28. ___$6___

29. ___14 cakes; 13 cakes will feed only 780 people.___

30. ___$66.15___

Answers

# Chapter 3 Answer Key

1. _____ 6 × 3 or 18 _____

2. _____ 6 × 11 or 66 _____

3. _____ 35 ÷ 7 or 5 _____

4. _____ 56 ÷ 8 or 7 _____

5. _____ 280 mi;
20 × 14 = 280 _____

6. _____ 7 bracelets;
49 ÷ 7 = 7 _____

7. _____ 9.6 mi _____

8. _____ $67.92 _____

9. _____ 2.72 m² _____

10. _____ 13.39 ft _____

11. _____ about $4.11 _____

12. _____ 68 _____

13. _____ 0.02492 _____

14. _____ 0.04 _____

15. _____ 750 _____

16. _____ 22 _____

17. _____ $5,950 _____

18. _____ $14.50 _____

19. _____ $3.33 _____

20. _____ 24 _____

21. _____ 7.04 _____

22. _____ 6.8 _____

23. _____ 0.515 _____

24. _____ 33 _____

25. _____ $28.99 _____

26. _____ 3.35 oz _____

27. _____ 0.315 mi _____

28. _____ $8 _____

29. _____ 14 cakes; 13 cakes will
feed only 715 people. _____

30. _____ $74.90 _____

**Course 1 • Chapter 3** Compute With Multi-Digit Numbers

# Chapter 4 Answer Key

**Are You Ready?—Review**
**Page 73**

1. 18

2. 7

3. 15

4. 9

5. 0

6. 6

7. 5

8. 4

**Are You Ready?—Practice**
**Page 74**

1. 13

2. 12

3. 19

4. 2

5. 6

6. about 4 inches

7. $1\frac{1}{3}$

8. $1\frac{4}{9}$

9. $1\frac{3}{16}$

10. $4\frac{1}{4}$

11. $5\frac{1}{10}$

12. $6\frac{2}{15}$

13. $5\frac{5}{6}$

14. $9\frac{1}{8}$ mi

Answers

# Chapter 4 Answer Key

**Are You Ready?—Apply**
**Page 75**

1. **MUSIC** Kristy downloads two songs to her MP3 player. The songs are $3\frac{3}{10}$ minutes and $4\frac{2}{3}$ minutes long. About how many minutes of memory will these two songs use altogether? **about 8 minutes**

2. **TRAINING** Mike ran $14\frac{1}{5}$ miles during his first week of training and $25\frac{3}{8}$ miles during his second week. About how many miles has he run in all? **about 39 miles**

3. **CATS** In a cat shelter, the longest cat is $18\frac{7}{8}$ inches long and the shortest cat is $13\frac{1}{4}$ inches long. About how much longer is the longest cat than the shortest cat? **about 6 in.**

4. **WALKING** Monette's piano instructor's house is $7\frac{1}{8}$ miles from her house. After biking for $1\frac{3}{4}$ miles, she stops to rest. How much longer does she need to bike to reach her instructor's house? $5\frac{3}{8}$ **miles**

5. **READING** Carole read $\frac{1}{8}$ of her book on Monday and $\frac{1}{6}$ of her book on Tuesday. What fraction of her book did she read altogether on these days? $\frac{7}{24}$

6. **ROPE** Mr. Silva cut $3\frac{9}{16}$ feet of rope from a bundle that was $10\frac{3}{4}$ feet long. What length of rope remains in the bundle? $7\frac{3}{16}$ **feet**

# Chapter 4 Answer Key

**Diagnostic Test**
**Page 76**

1. _____ 10 _____

2. _____ 12 _____

3. _____ 11 _____

4. _____ 1 _____

5. __ about 5 gallons __

6. _____ $\frac{5}{9}$ _____

7. _____ $\frac{7}{8}$ _____

8. _____ $\frac{3}{28}$ _____

9. _____ $2\frac{7}{10}$ _____

10. _____ $11\frac{1}{9}$ _____

11. _____ $5\frac{1}{2}$ _____

12. _____ $19\frac{11}{30}$ _____

13. __ $9\frac{9}{16}$ pounds __

**Pretest**
**Page 77**

1. _____ 5 _____

2. _____ 1 _____

3. _____ 40 _____

4. _____ 4 _____

5. _____ 12 _____

6. _____ $\frac{5}{24}$ _____

7. _____ $\frac{4}{9}$ _____

8. _____ 2 _____

9. _____ $\frac{1}{4}$ _____

10. _____ $\frac{5}{3}$ _____

11. _____ $7\frac{1}{2}$ _____

12. _____ $1\frac{3}{5}$ _____

13. _____ $\frac{1}{10}$ _____

14. _____ $8\frac{1}{10}$ _____

15. _____ $\frac{3}{4}$ _____

16. _____ $1\frac{2}{3}$ _____

Answers

# Chapter 4 Answer Key

Chapter Quiz
Page 78

1. $\frac{1}{3} \times 27 = 9$

2. $1 \times 2 = 2$

3. $4 \times 3 = 12$

4. $3 \times 3 = 9$

5. about 3 miles

6. 4

7. $2\frac{4}{7}$

8. 10

9. 9 muffins

10. 81ft

11. 27 flowers

Vocabulary Test
Page 79

1. fraction

2. improper fraction

3. numerator

4. Commutative Property

5. mixed number

6. denominator

7. unit ratio

8. reciprocal

9. Sample answer: process of including units of measurements as factors when you compute

10. Sample answer: the form of a fraction when the greatest common factor of the numerator and denominator is 1

# Chapter 4 Answer Key

**Student Recording Sheet, Page 82**

*Use this recording sheet with the Standardized Test Practice pages.*

**Fill in the correct answer. For gridded-response questions, write your answers in the boxes on the answer grid and fill in the bubbles to match your answers.**

1. Ⓐ Ⓑ ● Ⓓ

2. Ⓕ Ⓖ Ⓗ ●

3. ● Ⓑ Ⓒ Ⓓ

4. [gridded response: 1 2]

5. [gridded response: 4]

6. Ⓕ Ⓖ ● Ⓘ

7. ● Ⓑ Ⓒ Ⓓ

8. [gridded response: 1 / 4]

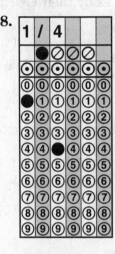

9. Ⓕ Ⓖ ● Ⓘ

10. Jordan biked $1\frac{1}{2}$ times more miles on Friday than Wednesday.

11. Ⓐ Ⓑ ● Ⓓ

12. Ⓕ ● Ⓗ Ⓘ

13. 9 c

## Extended Response

Record your answers for Exercise 14 on the back of this paper.

**Part A** Rose will need to multiply each ingredient by $2\frac{1}{2}$ in order to serve 30 people.

**Part B** She will need $3\frac{3}{4}$ c pretzels, $6\frac{7}{8}$ c dry cereal, $2\frac{13}{16}$ c peanuts, $\frac{5}{6}$ c margarine, and $\frac{25}{64}$ c soy sauce.

**Part C** $13\frac{7}{16}$ c; $3\frac{3}{4} + 6\frac{7}{8} + 2\frac{13}{16} = 13\frac{7}{16}$

Answers

# Chapter 4 Answer Key

*In addition to the scoring rubric, the following sample answers*
*may be used as guidance in evaluating open-ended assessment items.*

**1. a.** Estimate $4\frac{3}{4} \times 2\frac{2}{5}$. $4\frac{3}{4} \times 2\frac{2}{5}$ rounds to $5 \times 2$. So, an estimate is 10. You will need about 10 square feet of fabric.

    **b.** Find $4\frac{3}{4} \times 2\frac{2}{5}$. Write each mixed number as an improper fraction, $\frac{19}{4} \times \frac{12}{5}$. Then divide 4 and 12 by their GCF, 4. Multiply 19 and 3 then 1 and 5 to get $\frac{57}{5}$. Simplify to get $11\frac{2}{5}$.

**2. a.** Estimate $\frac{15}{16} \times \frac{1}{4}$. $\frac{15}{16}$ rounds to 1. $1 \times \frac{1}{4} = \frac{1}{4}$. So, an estimate is $\frac{1}{4}$. Each plastic rectangle is about $\frac{1}{4}$ in$^2$.

    **b.** $\frac{15}{16} \times \frac{1}{4} = \frac{15}{64}$. Each plastic rectangle is $\frac{15}{64}$ in$^2$.

**3.** Find how many groups of $\frac{3}{8}$ are in $4\frac{1}{2}$. Use division.

$$4\frac{1}{2} \div \frac{3}{8} = \frac{9}{2} \div \frac{3}{8} = \frac{9}{2} \times \frac{8}{3} = \frac{\overset{3}{\cancel{9}}}{\underset{1}{\cancel{2}}} \times \frac{\overset{4}{\cancel{8}}}{\underset{1}{\cancel{3}}} = \frac{12}{1} \text{ or } 12. \text{ The answer is 12 rectangles.}$$

# Chapter 4 Answer Key

1. ____**C**____

2. ____**G**____

3. ____**B**____

4. ____**G**____

5. ____**D**____

6. ____**F**____

7. ____**B**____

8. ____**F**____

9. ____**B**____

10. ____**H**____

11. ____**C**____

12. ____**H**____

13. ____**D**____

14. ____**G**____

15. ____**C**____

16. ____**G**____

17. ____**B**____

18. ____**I**____

19. ____**B**____

20. ____**H**____

**Answers**

# Chapter 4 Answer Key

1. ___C___

2. ___I___

3. ___B___

4. ___H___

5. ___D___

6. ___I___

7. ___A___

8. ___G___

9. ___A___

10. ___I___

11. ___B___

12. ___F___

13. ___D___

14. ___I___

15. ___B___

16. ___F___

17. ___D___

18. ___F___

19. ___B___

20. ___F___

# Chapter 4 Answer Key

**Test, Form 2A**
**Page 89**

1. ___B___

2. ___H___

3. ___D___

4. ___H___

5. ___A___

6. ___F___

7. ___C___

8. ___F___

9. ___D___

**Test, Form 2A** *(continued)*
**Page 90**

10. ___I___

11. ___A___

12. ___H___

13. ___A___

14. ___H___

15. ___5 rooms___

16. ___$1\frac{1}{3}$___

17. ___$3\frac{1}{11}$___

18. ___$5\frac{2}{3}$ ft___

19. ___$28\frac{4}{5}$ oz___

20. ___36 tiles___

**Answers**

# Chapter 4 Answer Key

1. _____ C _____

2. _____ G _____

3. _____ C _____

4. _____ I _____

5. _____ D _____

6. _____ G _____

7. _____ C _____

8. _____ F _____

9. _____ A _____

10. _____ F _____

11. _____ C _____

12. _____ H _____

13. _____ D _____

14. _____ G _____

15. __ **3 rooms** __

16. _____ 1 _____

17. __ $2\frac{10}{13}$ __

18. __ $4\frac{2}{3}$ ft __

19. __ **28 oz** __

20. __ **54 tiles** __

# Chapter 4 Answer Key

**Test, Form 3A**
**Page 93**

1. $\dfrac{1}{9} \times 36 = 4$

2. $1 \times \dfrac{1}{2} = \dfrac{1}{2}$

3. about 10 million pounds

4. $3\dfrac{5}{9}$

5. $\dfrac{5}{64}$

6. $1\dfrac{1}{8}$

7. $1\dfrac{1}{4}$ gal

8. $\dfrac{1}{16}$

9. 28 ft²

10. $20\dfrac{5}{8}$ dozen

11. 42 pages

**Test, Form 3A** *(continued)*
**Page 94**

12. 35

13. $\dfrac{9}{10}$

14. $\dfrac{14}{15}$

15. $\dfrac{1}{7}$

16. no; 29,040 ft $= 5\dfrac{1}{2}$ mi and $5\dfrac{1}{2}$ mi $< 6$ mi

17. 6 chores

18. 4 chains

19. $6\dfrac{3}{4}$ pt

20. 60 times

**Answers**

# Chapter 4 Answer Key

1. $\dfrac{1}{10} \times 50 = 5$

2. $\dfrac{1}{2} \times 2 = 1$

3. **about 390 mph**

4. $9\dfrac{9}{10}$

5. $\dfrac{3}{10}$

6. $1\dfrac{3}{16}$

7. $4\dfrac{1}{2}$ gal

8. $\dfrac{1}{12}$

9. $96

10. $28\dfrac{4}{5}$ min

11. **168 pages**

12. **63**

13. $\dfrac{5}{6}$

14. $6\dfrac{2}{3}$

15. $\dfrac{9}{64}$

16. **48 times**

17. **no; 30,360 ft $= 5\dfrac{3}{4}$ mi and $5\dfrac{3}{4}$ mi $< 6$ mi**

18. **4 pieces**

19. $7\dfrac{1}{4}$ pt

20. $12\dfrac{1}{3}$ times

# Chapter 5 Answer Key

**Are You Ready?—Review**
**Page 97**

1. $\dfrac{19}{20}$

2. $\dfrac{3}{20}$

3. $\dfrac{5}{9}$

4. $1\dfrac{3}{35}$

5. $1\dfrac{3}{8}$

6. $\dfrac{7}{12}$

7. $\dfrac{17}{45}$

8. $1\dfrac{41}{56}$

9. $\dfrac{41}{45}$

10. $\dfrac{13}{56}$

**Are You Ready?—Practice**
**Page 98**

1. $\dfrac{19}{24}$

2. $1\dfrac{7}{20}$

3. $\dfrac{1}{9}$

4. $\dfrac{16}{35}$

5. $1\dfrac{4}{45}$

6. $\dfrac{23}{35}$

7. $\dfrac{11}{40}$

8. $\dfrac{8}{35}$

9. $\dfrac{7}{12}$

10. $\dfrac{6}{35}$

11. $1\dfrac{3}{4}$

12. $\dfrac{8}{27}$

13. $1\dfrac{1}{24}$

14. $\dfrac{3}{8}$

Answers

# Chapter 5 Answer Key

**Are You Ready?—Apply**
**Page 99**

| | |
|---|---|
| **1. ATHLETICS** Two thirds of the track team ran laps after practice. One fourth of the team worked out in the weight room. What fraction more of the team ran laps than worked out in the weight room? $\dfrac{5}{12}$ | **2. ALLOWANCE** Eustace used $\dfrac{1}{2}$ of his allowance to buy CDs. He put $\dfrac{1}{3}$ of his allowance in savings. What fraction of his allowance did he use for CDs and savings? $\dfrac{5}{6}$ |
| **3. PAPER** Arcus had $\dfrac{8}{9}$ of a pack of construction paper. He used $\dfrac{1}{5}$ of the paper to make paper airplanes. What fraction of the original pack did Arcus use to make paper airplanes? $\dfrac{8}{45}$ | **4. MEASUREMENT** Lagan sold $\dfrac{7}{8}$ pound of chocolate for the school fundraiser. Nathan sold $\dfrac{6}{7}$ pound. How much more did Lagan sell than Nathan? $\dfrac{1}{56}$ **pound** |
| **5. CONSTRUCTION** A carpenter bought $\dfrac{1}{2}$ pound of size 8 nails and $\dfrac{3}{8}$ pound of size 16 nails. How many pounds of nails did he buy in all? $\dfrac{7}{8}$ **pound** | **6. FARMING** A farmer planted $\dfrac{4}{5}$ of a field in soybeans. One day, $\dfrac{5}{6}$ of the soybeans were picked. What fraction of the entire field was picked that day? $\dfrac{2}{3}$ |

# Chapter 5 Answer Key

**Diagnostic Test**
**Page 100**

1. _____ < _____

2. _____ > _____

3. _____ < _____

4. _____ < _____

5. _____ 214 < 241 _____

6. _____ 2,380 > 2,208 _____

7. _____ 0.875 _____

8. _____ 0.65 _____

9. _____ 0.16 _____

10. _____ 0.74 _____

**Pretest**
**Page 101**

1. _____ −$50 _____

2. _____ 145 _____

3. _____ 32 _____

4. _____ 14 _____

5. _____ −4 < 8 _____

6. _____ $0.\overline{4}$ _____

7. _____ $0.5\overline{3}$ _____

8. _____ $-0.91\overline{6}$ _____

9–12.

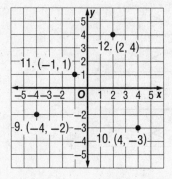

11. (−1, 1)
12. (2, 4)
9. (−4, −2)
10. (4, −3)

**Answers**

# Chapter 5 Answer Key

## Chapter Quiz
### Page 102

1. _____ −14 _____

2. _____ 7 _____

3. _____ −35 _____

4.

5.

6. _____ 46 _____

7. _____ 4 _____

8. _____ 17 _____

9. _____ 1,230 ft _____

10. _____ < _____

11. _____ > _____

12. _____ > _____

13. _____ > _____

14. _____ −10, −6, 3, 4, 7 _____

15. _____ −30, −24, 9, 16, 25 _____

16. _____ −101, −76, −48, 83, 93, 102 _____

## Vocabulary Test
### Page 103

1. positive integer

2. opposites

3. absolute value

4. rational number

5. bar notation

6. terminating decimal

7. repeating decimal

8. Sample answer: the regions a coordinate plane is separated into.

9. Sample answer: the set of positive whole numbers, their opposites, and zero.

**Course 1 · Chapter 5** Integers and the Coordinate Plane

# Chapter 5 Answer Key

*Use this recording sheet with the Standardized Test Practice pages.*

**Fill in the correct answer. For gridded-response questions, write your answers in the boxes on the answer grid and fill in the bubbles to match your answers.**

1. $\boxed{2\ 2\ /\ 6\ 7}$

2. ● Ⓑ Ⓒ Ⓓ

3. $\boxed{2\ 4}$

4. $\dfrac{4}{1}$

5. Ⓕ Ⓖ Ⓗ ●

6. Jacob; multiplying by a number less than 1 gives a product less than the other factor.

7. Ⓐ ● Ⓒ Ⓓ

8. 34 min

9. ● Ⓖ Ⓗ Ⓘ

10. Ⓐ Ⓑ Ⓒ ●

11. Ⓕ Ⓖ ● Ⓘ

12. Ⓐ ● Ⓒ Ⓓ

13. Ⓕ Ⓖ ● Ⓘ

## Extended Response

Record your answers for Exercise 14 on the back of this paper.

**Part A** $3.\overline{7}$

**Part B** $3.6, 3.65, 3.\overline{7}$

Answers

# Chapter 5 Answer Key

*In addition to the scoring rubric, the following sample answers may be used as guidance in evaluating extended response assessment items.*

1. **a.** $-\frac{7}{8}$; $-0.875 = -\frac{875}{1000} = -\frac{7}{8}$

   **b.** $-2\frac{5}{12} = -2.41\overline{6}$; using a calculator, $5 \div 12 = 0.41666667$.
   Use bar notation, $-2\frac{5}{12} = -2.41\overline{6}$.

   **c.** The elevation of the cave is greater than the elevation of the river, or the cave is higher than the river.

2. **a.** $(3, 4)$

   **b.** $(0, -2)$

   **c.** Fowls building

   **d.** Sample answer: $(-3, -1), (1, -1), (3, -1)$

# Chapter 5 Answer Key

1. _____ B _____

2. _____ H _____

3. _____ B _____

4. _____ H _____

5. _____ A _____

6. _____ I _____

7. _____ A _____

8. _____ G _____

9. _____ B _____

10. _____ G _____

11. _____ A _____

12. _____ F _____

13. _____ B _____

14. _____ G _____

15. _____ D _____

16. _____ H _____

17. _____ A _____

18. _____ G _____

19. _____ B _____

**Answers**

# Chapter 5 Answer Key

Test, Form 1B
Page 111

1. ___A___

2. ___H___

3. ___B___

4. ___G___

5. ___A___

6. ___G___

7. ___D___

8. ___F___

9. ___C___

Test, Form 1B  *(continued)*
Page 112

10. ___F___

11. ___C___

12. ___H___

13. ___B___

14. ___G___

15. ___D___

16. ___H___

17. ___B___

18. ___G___

19. ___D___

# Chapter 5 Answer Key

1. _____**C**_____

2. _____**H**_____

3. _____**B**_____

4. _____**I**_____

5. _____**B**_____

6. _____**G**_____

7. _____**C**_____

8. _____**G**_____

9. _____**A**_____

10. _____**I**_____

11. _____**C**_____

12. _____**H**_____

13. _____**B**_____

14. _____**F**_____

15–18.

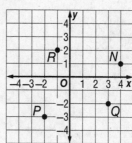

19. _____**>**_____

20. _____**<**_____

21. __**point *B***__

22. __**(1, 5)**__

23. __**(−4, −2)**__

24. __**(4, −2)**__

**Answers**

# Chapter 5 Answer Key

1. ___D___

2. ___G___

3. ___D___

4. ___I___

5. ___A___

6. ___F___

7. ___C___

8. ___F___

9. ___C___

10. ___F___

11. ___A___

12. ___F___

13. ___C___

14. ___G___

15–18.

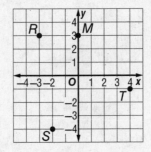

19. ___>___

20. ___>___

21. ___point *E*___

22. ___(−3, −4)___

23. ___(−3, 4)___

24. ___(3, 4)___

# Chapter 5 Answer Key

1. $-20$

2. $3$

3.
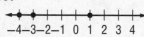

4. $75$

5. $10$

6. $3$

7. $1$

8. $23$

9. $-18$ feet

10.
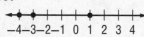

11. $>$

12. $>$

13. $<$

14. $>$

15. $=$

16. $-7, -5, -2, 0, 4$

17. $-0.\overline{63}$

18. $-0.72$

19. $-7\frac{4}{5}, -7.6, 7.\overline{34}, 7\frac{6}{11}$

20. $-10.\overline{20}, -10\frac{1}{5}, -10\frac{2}{11}, 10.19$

21.
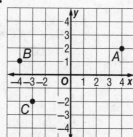

22. point $C$

23. $(-2, 3)$

24. $(-2, 0)$

25. Quadrant II

26. $(4, 2)$

27. $(4, 4)$

28. $(-1, -4)$

Answers

# Chapter 5 Answer Key

1.   5

2.   −4

3.
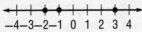

4.   45

5.   11

6.   4

7.   3

8.   28

9.   −13 feet

10.

11.   >

12.   >

13.   <

14.   =

15.   >

16.   −6, −4, −1, 0, 3

17.   $-0.4\overline{6}$

18.   −0.65

19.   $-5\frac{4}{5}$, −5.5, $5.\overline{33}$, $5\frac{5}{11}$

20.   $-8.\overline{21}$, $-8\frac{1}{5}$, $-8\frac{2}{11}$, 8.23

21.
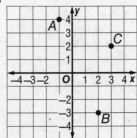

22.   point G

23.   (−4, −2)

24.   (2, −4)

25.   Quadrant III

26.   (1, 2)

27.   (1, 4)

28.   (−4, −2)

# Chapter 6 Answer Key

## Are You Ready?—Review
Page 121

1. _____ $9m^2$ _____

2. _____ $100\ in^2$ _____

3. _____ $16\ ft^2$ _____

4. _____ $121\ ft^2$ _____

5. _____ $49\ m^2$ _____

6. _____ $169\ cm^2$ _____

## Are You Ready?—Practice
Page 122

1. _____ $81\ ft^2$ _____

2. _____ $121\ cm^2$ _____

3. _____ $5\ m$ _____

4. _____ $1$ _____

5. _____ $\dfrac{3}{4}$ _____

6. _____ $\dfrac{4}{9}$ _____

7. _____ $1\dfrac{3}{7}$ _____

8. _____ $\dfrac{1}{6}$ _____

9. _____ $7\dfrac{11}{12}$ _____

10. _____ $4\dfrac{29}{30}$ _____

11. _____ $\dfrac{13}{40}$ _____

12. _____ $53\dfrac{1}{8}\ miles$ _____

**Answers**

# Chapter 6 Answer Key

| | |
|---|---|
| **1. GEOMETRY** Find the area of a square with a length of 6 inches.<br>**36 square inches** | **2. TABLECLOTHS** The area of a square tablecloth is 16 square feet. What is the length of the side of the tablecloth? **4 ft** |
| **3. BAKING** Katiana needs to bake a dozen muffins of a dozen different flavors for a school bake sale. How many muffins does Katiana need to bake? **144 muffins** | **4. WATER** Shasta poured water into three one-gallon water jugs to take to a race. She filled the first jug $\frac{3}{4}$ full. She filled each of the second and third jugs $\frac{7}{8}$ full. How much water did Shasta take to the race?<br>$2\frac{1}{2}$ **gallons** |
| **5. WALLPAPER** Yoki is putting up new wallpaper in her room. She wants to add a border along the ceiling. If her room is a rectangle with sides of $7\frac{1}{2}$ feet and $9\frac{3}{4}$ feet, how long of a border will she need?<br>$34\frac{1}{2}$ **feet** | **6. RESTAURANT** A restaurant sells pies by the slice. At the end of the night they have $\frac{1}{2}$ of a cherry pie, $\frac{2}{3}$ of an apple pie, and $\frac{1}{6}$ of a banana cream pie. How much total pie is left?<br>$1\frac{1}{3}$ **pies** |

# Chapter 6 Answer Key

**Diagnostic Test**
**Page 124**

1. _____ **196 cm²** _____

2. _____ **225 in²** _____

3. _____ **2 mi** _____

4. _____ $\dfrac{5}{8}$ _____

5. _____ $\dfrac{2}{5}$ _____

6. _____ $\dfrac{4}{9}$ _____

7. _____ $3\dfrac{15}{28}$ _____

8. _____ $4\dfrac{1}{30}$ _____

9. _____ $11\dfrac{3}{4}$ _____

10. _____ $\dfrac{3}{40}$ _____

11. _____ $3\dfrac{2}{3}$ **ft** _____

**Pretest**
**Page 125**

1. _____ **31** _____

2. _____ **7** _____

3. _____ **9** _____

4. _____ **20** _____

5. Let $a$ = Odile's age; $a + 9$.

6. Let $c$ = total cost; $c \div 2$.

7. Let $b$ = the number of berries; $3b$.

8. yes; Commutative Property

9. yes; Identity Property

10. _____ $6d + 42$ _____

11. _____ $2a + 20$ _____

12. _____ $5y - 25$ _____

Answers

# Chapter 6 Answer Key

**Chapter Quiz**
**Page 126**

1. $9^4$
2. $17^3$
3. $4^5$
4. $6^2$

5. $5 \times 5 \times 5$; 125
6. $4 \times 4 \times 4 \times 4 \times 4 \times 4$; 4,096
7. $7 \times 7 \times 7 \times 7$; 2,401
8. $11 \times 11$; 121

9. 6
10. 31
11. 13
12. 41
13. 16
14. 7
15. 43
16. 63
17. $4n$
18. $n \div 14$

**Vocabulary Test**
**Page 127**

1. Numerical expressions
2. order of operations
3. variable
4. Properties
5. Distributive Property
6. algebraic expression
7. Commutative
8. Associative
9. Sample answer: The number used as a factor.
10. Sample answer: Tells how many times the base is used as a factor.

**Student Recording Sheet, Page 130**

*Use this recording sheet with the Standardized Test Practice pages.*

**Fill in the correct answer. For gridded-response questions, write your answers in the boxes on the answer grid and fill in the bubbles to match your answers.**

1. Ⓐ Ⓑ Ⓒ ●

2. Ⓕ ● Ⓗ Ⓘ

3. Ⓐ Ⓑ Ⓒ ●

4. **2 2 . 9 4**

5. Ⓕ Ⓖ Ⓗ ●

6. _____ 4 _____

7. Ⓐ ● Ⓒ Ⓓ

8. Ⓕ ● Ⓗ Ⓘ

9. ● Ⓑ Ⓒ Ⓓ

10. ● Ⓖ Ⓗ Ⓘ

11. **0 . 3**

12. Ⓐ ● Ⓒ Ⓓ

13. **1 0**

14a. $5h - 6$

14b. $9; 5(3) - 6 = 15 - 6 = 9$

## Extended Response

**Record your answers for Exercise 14 on the back of this paper.**

Answers

# Chapter 6 Answer Key

## Extended-Response Test, Page 131
## Sample Answers

*In addition to the scoring rubric, the following sample answers may be used as guidance in evaluating extended response assessment items.*

1. Sample answer: First, evaluate anything in parentheses using the remaining order of operations. Then, evaluate the exponents. After that, evaluate any multiplication and division in order from left to right. Last, evaluate any addition and subtraction in order from left to right.

2. **a.** $15 + s$

   **b.** $15 + 25i$

   **c.** Sample answer: Yes. The value of $s$ in the first expression is the same as the value of $25i$ in the second expression. Both are being added to 15, so both expressions have the same value.

3. $7(6) + 6(6 \div 3)^2 - 2(3)$       Substitute 6 for $p$ and 3 for $q$.

   $7(6) + 6(2)^2 - 2(3)$          Evaluate inside the parentheses.

   $7(6) + 6(4) - 2(3)$             Evaluate the exponent.

   $42 + 24 - 6$                    Multiply.

   $60$                             Add and subtract.

4. Sample answer: Using the Commutative Property, we can rewrite the first expression as $(b + a) + c$. Then we can use the Associative Property to rewrite it as $b + (a + c)$.

# Chapter 6 Answer Key

1. _____ D _____

2. _____ F _____

3. _____ B _____

4. _____ F _____

5. _____ B _____

6. _____ G _____

7. _____ C _____

8. _____ G _____

9. _____ A _____

10. _____ G _____

11. _____ D _____

12. _____ H _____

13. _____ D _____

14. _____ G _____

15. _____ C _____

16. _____ H _____

17. _____ D _____

18. _____ I _____

19. _____ B _____

20. _____ I _____

21. _____ C _____

22. _____ I _____

23. _____ A _____

24. _____ H _____

25. _____ B _____

**Answers**

# Chapter 6 Answer Key

**Test, Form 1B**
**Page 135**

1. A
2. G
3. C
4. G
5. B
6. H
7. B
8. H
9. C
10. F
11. A
12. I

**Test, Form 1B** ***(continued)***
**Page 136**

13. B
14. I
15. B
16. G
17. B
18. F
19. D
20. H
21. A
22. F
23. B
24. F
25. A

# Chapter 6 Answer Key

1. _____ C _____

2. _____ I _____

3. _____ C _____

4. _____ H _____

5. _____ A _____

6. _____ G _____

7. _____ D _____

8. _____ G _____

9. _____ B _____

10. _____ G _____

11. _____ D _____

12. _____ F _____

13. _____ C _____

14. _____ I _____

15. _____ A _____

16. _____ G _____

17. _____ C _____

18. _____ $3x + 36$ _____

19. _____ $40 + 5r$ _____

20. _____ 45 _____

21. _____ 20 _____

22. _____ yes; Associative _____

23. _____ yes; Commutative _____

24. _____ $14x$ _____

25. _____ $14x$ _____

26. _____ $32x + 12y$ _____

27. _____ $3(x + 3) + 3x$; $6x + 9$ _____

**Answers**

# Chapter 6 Answer Key

1. _____A_____

2. _____F_____

3. _____B_____

4. _____G_____

5. _____C_____

6. _____H_____

7. _____A_____

8. _____G_____

9. _____A_____

10. _____H_____

11. _____D_____

12. _____G_____

13. _____B_____

14. _____F_____

15. _____B_____

16. _____I_____

17. _____D_____

18. _____$4y + 32$_____

19. _____$6 + 3h$_____

20. _____30_____

21. _____54_____

22. _____yes; Associative_____

23. _____yes; Commutative_____

24. _____18x_____

25. _____45x_____

26. _____$21x + 6y$_____

27. _____$4(x + 2) + 2x$; $6x + $8_____

# Chapter 6 Answer Key

1.  $\frac{1}{6} \times \frac{1}{6} \times \frac{1}{6}$; $\frac{1}{216}$

2.  $9^6$

3.  18

4.  221

5.  100

6.  22

7.  4($11.79) + 7($6.65); $93.71

8.  11

9.  21

10.  42

11.  $13.75

12.  $n + 16$

13.  $2p + 1$

14.  $d \div 3$

15.  $4w - 6$

16.  $8n + 3$

17.  $p \div 4$

18.  yes; Associative

19.  yes; Identity

20.  no; $12 - (5 - 3) = 10$ and $(12 - 5) - 3 = 4$

21.  $45 + 22 + 25 =$ $45 + 25 + 22 =$ $70 + 22 = 92$ minutes

22.  $6(50) + 6(2) =$ $300 + 12 = 312$

23.  $4(7) + 4(0.1) =$ $28 + 0.4 = 28.4$

24.  $5x + 45$

25.  $132 + 11r$

26.  $9v + 18$

27.  $7b + 24.5$

28.  $14($12) + 14($4.95) =$ $14($16.95) = $237.30

29.  $14x$

30.  $10x + 20y$

31.  $15x + 2y$

32.  $9(x + 4y)$

33.  $6(2x + 3y)$

Answers

# Chapter 6 Answer Key

1. $\frac{1}{9} \times \frac{1}{9} \times \frac{1}{9}; \frac{1}{729}$

2. $4^8$

3. $10$

4. $237$

5. $39$

6. $15$

7. $5(\$5.35) + 9(\$3.45);$ $\$57.80$

8. $19$

9. $20$

10. $4\frac{3}{5}$

11. $\$8.75$

12. $s \div 4$

13. $t - 30$

14. $2p + 3$

15. $3h - 8$

16. $3r + 40$

17. $g \div 5$

18. yes; Identity

19. yes; Commutative

20. no; $10 \div (8 \div 4) = 5$ and $(10 \div 8) \div 4 = 0.3125$

21. $18 + 26 + 14 =$ $18 + (26 + 14) =$ $18 + 40 = 58$ minutes

22. $9(30) + 9(4) =$ $270 + 36 = 306$

23. $4(8) + 4(0.2) =$ $32 + 0.8 = 32.8$

24. $3y + 30$

25. $182 + 13t$

26. $8a + 32$

27. $3w + 4.8$

28. $12(\$6.49) + 12(\$8) =$ $12(\$14.49) = \$173.88$

29. $16x$

30. $12x + 28y$

31. $5x + 20y$

32. $6(7x + 2y)$

33. $15(x + 2y)$

# Chapter 7 Answer Key

**Are You Ready?—Review**
Page 145

1. $\dfrac{1}{6}$

2. $\dfrac{5}{8}$

3. $\dfrac{3}{10}$

4. $\dfrac{3}{40}$ mi

5. $\dfrac{4}{9}$

6. $\dfrac{2}{21}$

7. $\dfrac{1}{5}$

8. $\dfrac{1}{8}$ gal

**Are You Ready?—Practice**
Page 146

1. 3.36

2. 0.3

3. 1.55

4. 1.37

5. $1.75

6. $\dfrac{1}{2}$

7. $\dfrac{11}{36}$

8. $\dfrac{9}{20}$

9. $\dfrac{3}{10}$

10. $\dfrac{1}{6}$ min

11. $\dfrac{17}{24}$ gallon

**Answers**

# Chapter 7 Answer Key

1. **FARMING** A farmer plows $\frac{2}{3}$ of his field in the morning and then $\frac{1}{6}$ of his field in the afternoon. How much more of the field did he mow in the morning?

   $\frac{1}{2}$ **of the field**

2. **TOURING** On a guided tour of a cave, a guide walks for a total of $\frac{3}{4}$ of an hour and stops and talks for a total of $\frac{2}{5}$ of an hour. How much longer was the walking portion than the talking portion of the tour?

   $\frac{7}{20}$ **h or 21 min**

3. **CHEF HAT** Brynn made a chef's hat for her father. The top of the hat was made with $\frac{6}{7}$ yard of linen and the brim of the hat was made with $\frac{2}{5}$ yard of linen. How much more linen was used for the top of the hat than the brim?

   $\frac{16}{35}$ **yd**

4. **BIRD FEEDER** A bird feeder held $\frac{7}{8}$ cup of birdseed until a flock of Northern Mockingbirds ate $\frac{1}{2}$ cup of seed. How much seed was left in the birdfeeder after that? $\frac{3}{8}$ **c**

5. **WALK-A-THON** After one hour of walking, Sayaka had $\frac{4}{5}$ of the walk completed while Takara had $\frac{7}{10}$ of the walk completed. How much more of the walk had Sayaka completed than Takara? $\frac{1}{10}$ **of the walk**

6. **COOK-OFF** For a dip recipe, Atrina used $\frac{5}{6}$ of a pound of cream cheese. Kendra used $\frac{2}{3}$ of a pound of cream cheese. How much less cream cheese did Kendra use? $\frac{1}{6}$ **lb**

# Chapter 7 Answer Key

**Diagnostic Test**
**Page 148**

1. _____ **3.53** _____

2. _____ **0.26** _____

3. _____ **1.12** _____

4. _____ **3.28** _____

5. _____ **$0.45** _____

6. _____ $\dfrac{5}{24}$ _____

7. _____ $\dfrac{3}{8}$ _____

8. _____ $\dfrac{1}{3}$ _____

9. _____ $\dfrac{1}{15}$ _____

10. $\dfrac{1}{15}$ **more of the group**

11. _____ $\dfrac{1}{40}$ **mi** _____

**Pretest**
**Page 149**

1. _____ **6** _____

2. _____ **10** _____

3. _____ **9** _____

4. _____ **14** _____

5. _____ **4** _____

6. _____ **13** _____

7. _____ **6** _____

8. _____ **50** _____

9. _____ **45** _____

10. $n + 12 = 24;$
    **12 in.**

11. $w - 1 = 9;$
    **10 words**

12. _____ **$19** _____

**Answers**

# Chapter 7 Answer Key

## Chapter Quiz
### Page 150

1. _____4_____

2. _____7_____

3. _____8_____

4. _____9_____

5. _____28_____

6. _____$9_____

7. _____11_____

8. _____21.5_____

9. _____39_____

10. $f + 12 = 35$; 23 cards

11. Sample answer: 9, 2, and 6

## Vocabulary Test
### Page 151

1. _____equals sign_____

2. _____solution_____

3. _____Division_____

4. _____inverse operations_____

5. _____Addition_____

6. _____equation_____

7. _____inverse operations_____

8. You can subtract the same amount from each side of an equation and the equation remains true.

9. To replace a variable in an equation with a value that makes it true

# Chapter 7 Answer Key

**Student Recording Sheet, Page 154**

*Use this recording sheet with the Standardized Test Practice pages.*

**Fill in the correct answer. For gridded-response questions, write your answers in the boxes on the answer grid and fill in the bubbles to match your answers.**

1. Ⓐ Ⓑ ● Ⓓ

2. Ⓕ ● Ⓗ Ⓘ

3.
```
2 / 5
```

4. ● Ⓑ Ⓒ Ⓓ

5.
```
0 . 2 8
```

6. Ⓕ Ⓖ Ⓗ ●

7.
```
9 3 6
```

8. Ⓐ Ⓑ Ⓒ ●

9. ___**15 seats**___

10. Ⓕ Ⓖ ● Ⓘ

11. Ⓐ ● Ⓒ Ⓓ

12. ___$8h = 96$; $12h$___

13a. Fisherman's service, $480

13b. Wild Fishing Inc.

13c. Sample answer: Let $x$ represent the number of days Wild Fishing Inc: $400 + 50x$; Fisherman's service: $300 + 60x$

# Extended Response
Record your answers for Exercise 13 on the back of this paper.

**Answers**

# Chapter 7 Answer Key
## Extended-Response Test, Page 155
## Sample Answers

*In addition to the scoring rubric, the following sample answers may be used as guidance in evaluating extended response assessment items.*

1. $5m = 975$; $m = 195$; The cost is $195 per student.

2. $x + 10.50 = 38.50$; $x = 28$; The cost of a bear is $28.

3. $12.50m = 175$; $m = 14$; The leader bought 14 cheerleading outfits.

4. $9.50n = 57$; $n = 6$; She bought 6 soccer uniforms.

# Chapter 7 Answer Key

**Test, Form 1A**
**Page 157**

1. _____ **B** _____

2. _____ **I** _____

3. _____ **C** _____

4. _____ **I** _____

5. _____ **D** _____

6. _____ **G** _____

7. _____ **A** _____

8. _____ **H** _____

9. _____ **A** _____

10. _____ **G** _____

**Test, Form 1A** *(continued)*
**Page 158**

11. _____ **A** _____

12. _____ **I** _____

13. _____ **C** _____

14. _____ **G** _____

15. _____ **C** _____

16. _____ **F** _____

17. _____ **C** _____

18. _____ **H** _____

19. _____ **D** _____

20. _____ **H** _____

**Answers**

# Chapter 7 Answer Key

1. ___C___

2. ___H___

3. ___B___

4. ___I___

5. ___D___

6. ___H___

7. ___C___

8. ___G___

9. ___C___

10. ___H___

11. ___A___

12. ___F___

13. ___D___

14. ___H___

15. ___D___

16. ___H___

17. ___D___

18. ___I___

19. ___A___

20. ___H___

# Chapter 7 Answer Key

**Test, Form 2A**
**Page 161**

1. _____ **C** _____

2. _____ **F** _____

3. _____ **D** _____

4. _____ **I** _____

5. _____ **D** _____

6. _____ **I** _____

7. _____ **C** _____

8. _____ **H** _____

9. _____ **D** _____

**Test, Form 2A** *(continued)*
**Page 162**

10. _____ **F** _____

11. _____ **D** _____

12. _____ **G** _____

13. _____ **C** _____

14. _____ **G** _____

15. _____ **B** _____

16. _____ **242** _____

17. _____ **0.5** _____

18. _____ **294** _____

19. _____ $50q = 350$; **7 quarters** _____

20. _____ **9 rooms** _____

**Answers**

# Chapter 7 Answer Key

**Test, Form 2B**
Page 163

1. ___B___

2. ___I___

3. ___A___

4. ___H___

5. ___A___

6. ___F___

7. ___D___

8. ___G___

9. ___A___

**Test, Form 2B** *(continued)*
Page 164

10. ___H___

11. ___C___

12. ___G___

13. ___A___

14. ___I___

15. ___A___

16. ___189___

17. ___0.25___

18. ___384___

19. ___$15d = 135$; 9 dimes___

20. ___6 rooms___

# Chapter 7 Answer Key

**Test, Form 3A**
**Page 165**

1. ___25___

2. ___45___

3. ___4___

4. **59 shells**

5. ___26___

6. ___19.7___

7. $\dfrac{2}{9}$

8. ___136___

9. $a + 25 = 56$;
   **31 apples**

10. $p - 25 = 95$;
    **120 pages**

11. $h - 64 = 26$;
    **90°F**

**Test, Form 3A**  *(continued)*
**Page 166**

12. ___18___

13. ___13___

14. $\dfrac{3}{7}$

15. ___484___

16. ___116.2___

17. ___45___

18. ___3___

19. ___212___

20. $1.25r = 22.5$;
    **18 rides**

21. $\dfrac{s}{4} = 196$;
    **784 students**

22. $3x = 120$;
    **40 minutes**

23. **16 color copies**

**Answers**

# Chapter 7 Answer Key

1. 26

2. 60

3. 3

4. 58 markers

5. 28

6. 19.1

7. $\dfrac{3}{8}$

8. 121

9. $h + 11 = 96$;
   85 points

10. $m - 47 = 6$;
    $53

11. $h - 4 = 28$;
    32 lb

12. 12

13. 16

14. $\dfrac{5}{6}$

15. 624

16. 91.2

17. 40

18. 5

19. 162

20. $4.25h = 38.25$;
    9 hours

21. $\dfrac{s}{3} = 214$;
    642 students

22. $2x = 70$;
    35 minutes

23. 8 roses

# Chapter 8 Answer Key

**Are You Ready?—Review**
Page 169

**Are You Ready?—Practice**
Page 170

1. < 

2. < 

3. > 

4. < 

5. $408 < $480

6. 1,324 < 1,423

7. 42

8. 16

9. 27

10. 19

1. 35

2. 28

3. 16

4. 8

5. 3

6. 2

7. 25 lb

8. 40 gal

11. $60

12. 5

13. 6

14. 5

15. 6

16. 20 min

# Chapter 8 Answer Key

1. **ARCHERY** The school wants to buy new bows for the archery team. One store charges $485 per bow. Another store charges $505. How much does the cheaper bow cost? **$485**

2. **BICYCLING** Adam rode his bike every day after school for one week. He rode 80 minutes in all. How much time did he spend riding daily if he rode the same amount of time each day? **16 min**

3. **SCHOOL** The school collected donations for new equipment for three days. A total of $590 was donated. How much was donated on the second day? **$165**

| Day | Donations ($) |
|-----|---------------|
| 1   | 145           |
| 2   |               |
| 3   | 280           |

4. **MONEY** An investor made $2,390 on Stock A. He made $2,250 on Stock B. Which stock made more money? **Stock A**

5. **SONGS** The school choir practices 4 songs for 60 minutes. How much time do they spend practicing each song if they spend the same amount of time on each? **15 min**

6. **BASKETBALL** Basketball practice at the school gym lasts for 45 minutes. There are 15 students on the basketball team. How much time does the coach spend helping each member if he spends the same amount of time with each? **3 min**

# Chapter 8 Answer Key

**Diagnostic Test**
**Page 172**

1. _____ < _____

2. _____ > _____

3. _____ < _____

4. _____ < _____

5. _____ 214 < 241 _____

6. _____ 2,380 > 2,208 _____

7. _____ 34 _____

8. _____ 18 _____

9. _____ 25 _____

10. _____ 18 _____

11. _____ 74 cars _____

12. _____ 3 _____

13. _____ 7 _____

14. _____ 2 _____

15. _____ 8 _____

16. _____ 15 min _____

**Pretest**
**Page 173**

1.

| Input (x) | x + 10 | Output (y) |
|-----------|--------|------------|
| 5 | 5 + 10 | 15 |
| 8 | 8 + 10 | 18 |
| 11 | 11 + 10 | 21 |

2. _____ 13 _____

3.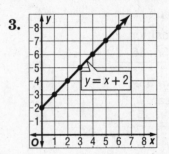

$y = x + 2$

4. _____ 6p _____

5. _____ yes _____

6. _____ no _____

7. _____ yes _____

8.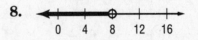

9. _____ $x \geq 6$ _____

10. _____ $x \leq 12$ _____

# Chapter 8 Answer Key

## Chapter Quiz
### Page 174

**1.**

| Input, $x$ | $x - 4$ | Output, $y$ |
|:---:|:---:|:---:|
| 4 | $4 - 4$ | 0 |
| 6 | $6 - 4$ | 2 |
| 8 | $8 - 4$ | 4 |

**2.** Multiply the position number by 3; $3n$; 36

**3.** Subtract 5 from the position number; $n - 5$; 7

**4.** $y = 4x$

**5.** $y = 12x$

**6.** $p = 6t$

**7.**

| Touchdowns, $t$ | 1 | 2 | 3 |
|:---:|:---:|:---:|:---:|
| Points, $p$ | 6 | 12 | 18 |

**8.**

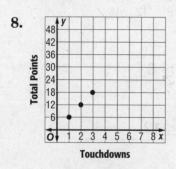

**9.** 42 points

## Vocabulary Test
### Page 175

**1.** sequence

**2.** independent

**3.** linear

**4.** function table

**5.** term

**6.** function

**7.** dependent

**8.** Sample answer: An inequality says that two expressions are *not* equal.

**9.** Sample answer: In an arithmetic sequence, each term is found by adding the same amount to the previous term.

# Chapter 8 Answer Key

*Use this recording sheet with the Standardized Test Practice pages.*

**Fill in the correct answer. For gridded-response questions, write your answers in the boxes on the answer grid and fill in the bubbles to match your answers.**

1. Ⓐ Ⓑ ● Ⓓ

2. ● Ⓖ Ⓗ Ⓘ

3.

4. Ⓐ ● Ⓒ Ⓓ

5.

6. Ⓕ ● Ⓗ Ⓘ

7. ● Ⓑ Ⓒ Ⓓ

8. $4$ to $1$, $4:1$, or $\frac{4}{1}$

9. Ⓕ ● Ⓗ Ⓘ

10. Jacob; Multiplying by a decimal less than 1 is the same as taking a portion of the whole number.

11. Ⓐ Ⓑ Ⓒ ●

12. 34 min

13A.

| Miles Driven | Cost ($) |
| --- | --- |
| 0 | 25 |
| 100 | 35 |
| 200 | 45 |
| 300 | 55 |
| 400 | 65 |
| 500 | 75 |

13B. $(x \div 10) + 25$

13C. $65

13D. For every 100 miles, driven the cost increases by $10, after an initial change of $25.

## Extended Response
**Record your answers for Exercise 13 on the back of this paper.**

Answers

# Chapter 8 Answer Key

**Extended-Response Test, Page 179**
**Sample Answers**

*In addition to the scoring rubric, the following sample answers may be used as guidance in evaluating extended response assessment items.*

**1. a–b.**

| Input ($x$) | Output ($3x - 1$) |
|:---:|:---:|
| 1 | 2 |
| 2 | 5 |
| 3 | 8 |
| 4 | 11 |
| 6 | 17 |
| 10 | 29 |

**2. a.** Multiply the number of tickets purchased by 7 to find the total cost.

 **b.** $7x$, where $x$ is the number of tickets purchased

 **c.** $y = 7x$; $y$ = total cost and $x$ = number of tickets purchased

 **d.** $y = 7x$
 $y = 7 \times 7$
 $y = \$49$
 It will cost a family $49 to buy 7 tickets.

 **e.** $7x \geq 420$
 $\dfrac{7x}{7} \geq \dfrac{420}{7}$
 $x \geq 60$
 So, the sixth grade class bought at least 60 tickets.

# Chapter 8 Answer Key

**Test, Form 1A**
**Page 181**

1. _____C_____

2. _____H_____

3. _____C_____

4. _____I_____

5. _____C_____

6. _____H_____

**Test, Form 1A** *(continued)*
**Page 182**

7. _____A_____

8. _____I_____

9. _____C_____

10. _____I_____

11. _____C_____

12. _____G_____

13. _____D_____

14. _____H_____

15. _____A_____

16. _____G_____

17. _____C_____

**Answers**

# Chapter 8 Answer Key

1. ___B___

2. ___F___

3. ___B___

4. ___I___

5. ___A___

6. ___G___

7. ___D___

8. ___H___

9. ___B___

10. ___F___

11. ___C___

12. ___H___

13. ___C___

14. ___G___

15. ___A___

16. ___F___

17. ___B___

**Course 1 • Chapter 8** Functions and Inequalities

# Chapter 8 Answer Key

**Test, Form 2A**
**Page 185**

1. ___D___

2. ___G___

3. ___C___

4. ___H___

5. ___D___

6. ___G___

7. ___D___

8. ___F___

9. ___C___

**Test, Form 2A** *(continued)*
**Page 186**

10. ___x + 2___

11. ___3x___

12. ___x ÷ 4___

13. Subtract 6 from the position number; $n - 6$

14. ___10___

15. $y = 10x + 25$

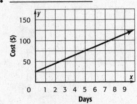

16. $c = 15d - 5$; $85

17. $y = 4x$

18. $y = 3x - 1$

19. ___14___

20. $p \le 45$

43  44  45  46  47

21. $x < 23$

21  22  23  24  25

22. $n \ge 18$

16  17  18  19  20

Answers

# Chapter 8 Answer Key

1. ___B___

2. ___G___

3. ___C___

4. ___H___

5. ___D___

6. ___I___

7. ___D___

8. ___F___

9. ___C___

10. ___4x___

11. ___$x \div 3$___

12. ___$x - 2$___

13. ___multiply the position number by 8; 8n___

14. ___120___

15. ___$y = 20x + 35$___

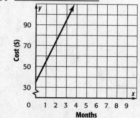

16. ___$c = 9d - 6$; $66___

17. ___$y = 3x$___

18. ___$y = 4x - 2$___

19. ___11___

20. ___$h \leq 10$___

8  9  10  11  12

21. ___$x \geq 16$___

14  15  16  17  18

22. ___$x < 24$___

22  23  24  25  26

# Chapter 8 Answer Key

1. _____6_____

| Input (x) | 3x − 2 | Output (y) |
|---|---|---|
| 3 | 3(3) − 2 | 7 |
| 6 | 3(6) − 2 | 16 |
| 8 | 3(8) − 2 | 22 |

2. _____

3. _____3x_____

4. _____x ÷ 4_____

5. _____2x + 3_____

6. _____a > 14_____

   12  13  14  15  16

7. _____s ≤ 8_____

   6   7   8   9   10

8. _____a < 6_____

   4   5   6   7   8

9. _____c ≥ 22_____

   20  21  22  23  24

10. _____m ≤ 6_____

   4   5   6   7   8

11. _____h > 45_____

   43  44  45  46  47

12. __s + 439 ≤ 1,250; s ≤ 811__

13. __multiply the position number by 9; 9n__

14. _____144_____

15. _____y = 5x_____

16. _____y = 7x + 2_____

17. _____p = 40d_____

| Days, d | 1 | 2 | 3 |
|---|---|---|---|
| Pounds, p | 40 | 80 | 120 |

18.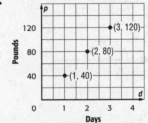

19. __The graph is a line because each day the amount increases by 40.__

20.

21.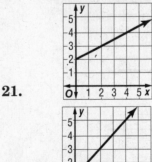

22.

**Answers**

# Chapter 8 Answer Key

1. _____ 6 _____

| Input (x) | 2x + 4 | Output (y) |
|-----------|----------|------------|
| 2 | 2(2) + 4 | 8 |
| 5 | 2(5) + 4 | 14 |
| 10 | 2(10) + 4 | 24 |

2. _____

3. _____ $x \div 3$ _____

4. _____ $3x - 1$ _____

5. _____ $2x$ _____

6. _____ $d > 12$ _____

7. _____ $a \le 6$ _____

8. _____ $m > 33$ _____

9. _____ $d \le 10$ _____

10. _____ $x \ge 15$ _____

11. _____ $n < 36$ _____

12. _____ $s + 358 \le 1{,}250$;
_____ $s \le 892$ _____

13. _____ multiply the position number by 8; 8n _____

14. _____ **128** _____

15. _____ $y = 5x + 1$ _____

16. _____ $y = 3x - 2$ _____

17. _____ $p = 75d$ _____

| Days, d | 1 | 2 | 3 |
|---------|-----|-----|-----|
| Pounds, p | 75 | 150 | 225 |

18.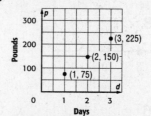

19. _____

The graph is a line because each day the amount increases by 40.

20. _____

21.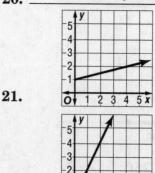

22.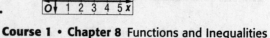

# Chapter 9 Answer Key

**Are You Ready?—Review**
**Page 193**

1. _____ 1 _____

2. _____ 2 _____

3. _____ 7 _____

4. _____ 12 _____

5. _____ 6 cm² _____

6. _____ 88 in² _____

**Are You Ready?—Practice**
**Page 194**

1. _____ 1 _____

2. _____ 4 _____

3. _____ 7 _____

4. _____ 6 _____

5. _____ 22 _____

6. _____ 11 _____

7. _____ 30 _____

8. _____ 21 cm² _____

9. _____ 40 ft² _____

10. _____ 6 yd² _____

**Answers**

# Chapter 9 Answer Key

1. **FARMING** A farmer has a field that is 40 yards long by 20 yards wide. What is the total area of the field? **800 yd²**

2. **WALLPAPER** Sydney is going to wallpaper her bedroom wall. If her wall is 10 feet long by 8 feet tall, how much wallpaper will she need? **80 ft²**

3. **GARDENING** Elena wants to put a new layer of soil in her garden. She needs to know the area of her garden so that she knows how much soil to buy. If the garden is 14 meters by 6 meters, how much area does she need to cover? **84 m²**

4. **TILES** Mr. McCabe is buying tiles for his kitchen floor. The floor is 12 feet long by 11 feet wide. If each tile is 1 square foot, how many tiles does he need? **132 tiles**

5. **STORM** A weather station reported that a strong storm covered an area of land that was 13 kilometers long by 15 kilometers wide. How much land area was covered by the storm? **195 km²**

6. **WILDFIRES** A spokesperson from the forest fire service said that a wildfire caused a patch of forest that was 20 miles long by 20 miles wide to be burned. How many square miles of forest were burned by the wildfire? **400 mi²**

# Chapter 9 Answer Key

**Diagnostic Test**
**Page 196**

1. _____ 2

2. _____ 15

3. _____ 3

4. _____ 12

5. _____ 9

6. _____ 12

7. _____ 40

8. _____ 56 cm$^2$

9. _____ 12 ft$^2$

10. _____ 144 yd$^2$

**Pretest**
**Page 197**

1. _____ 24 cm$^2$

2. _____ 30 ft$^2$

3. _____ 112 m$^2$

4. _____ 12 in$^2$

5. _____ 40 in$^2$

6. _____ 84 m$^2$

7. _____ 144 ft$^2$

Answers

# Chapter 9 Answer Key

1. ___108 cm²___

2. ___12.6 m²___

3. ___37.5 ft²___

4. ___60 in²___

5. ___32 in²___

6. ___14 m___

7. ___35 in.___

8. ___20 yd___

1. ___composite figure___

2. ___polygon___

3. ___base___

4. ___parallelogram___

5. ___equation___

6. ___height___

7. ___Sample answer: a parallelogram with 4 equal sides___

8. ___Sample answer: having the same measure___

# Chapter 9 Answer Key

**Student Recording Sheet, Page 202**

*Use this recording sheet with the Standardized Test Practice pages.*

**Fill in the correct answer. For gridded-response questions, write your answers in the boxes on the answer grid and fill in the bubbles to match your answers.**

1. Ⓐ ● Ⓒ Ⓓ

2.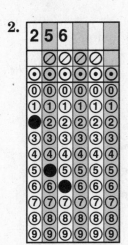

3. Ⓕ Ⓖ ● Ⓘ

4. Ⓐ ● Ⓒ Ⓓ

5. _____ 3 to 4 _____

6. Ⓕ ● Ⓗ Ⓘ

7. Ⓐ Ⓑ Ⓒ ●

8. _____ 105 ft² _____

9. Ⓕ Ⓖ Ⓗ ●

10. Ⓐ ● Ⓒ Ⓓ

11.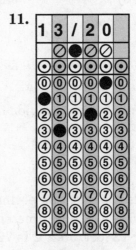

12. Ⓕ Ⓖ Ⓗ ●

13a. 32 ft

13b. It doubles. The perimeter of the new figure is 64 feet. This is equal to 32 times 2.

## Extended Response

Record your answers for Exercise 13 on the back of this paper.

Answers

# Chapter 9 Answer Key

### Extended-Response Test, Page 203
### Sample Answers

*In addition to the scoring rubric, the following sample answers may be used as guidance in evaluating extended response assessment items.*

1.  **a.** She needs 15 in² fabric. You can find the area of the triangle by using $A = \frac{1}{2}bh$. Multiply the base 5 by the height 6 and divide by 2.

    **b.** She needs 375 in² fabric. You can find the area of the parallelogram by using $A = bh$. Multiply the base 25 by the height 15.

    **c.** She needs 115 in² fabric. You can find the area of the trapezoid by using $A = \frac{1}{2}h(b_1 + b_2)$. Add the bases, 14 and 9. Then multiply by the height, 10. Divide by 2.

2.  To find the perimeter, add the sides to get 36 ft. To find the area, use the formula for the area of a trapezoid, $A = \frac{1}{2}h(b_1 + b_2)$. Add the bases, 6 and 14. Then multiply by 6 and divide by 2. The area of the trapezoid is 60 ft².

3.  The perimeter will also double. The perimeter of the original figure is 6 + 6 + 14 + 10 or 36 feet. The perimeter of the new figure is 12 + 12 + 28 + 20 or 72 feet. 36 × 2 is 72.

# Chapter 9 Answer Key

1. ___D___

2. ___G___

3. ___C___

4. ___H___

5. ___B___

6. ___I___

7. ___B___

8. ___I___

9. ___D___

10. ___F___

11. ___C___

Answers

# Chapter 9 Answer Key

Test, Form 1B
Page 207

Test, Form 1B *(continued)*
Page 208

1. ___A___

2. ___I___

3. ___B___

4. ___G___

5. ___C___

6. ___I___

7. ___B___

8. ___I___

9. ___D___

10. ___F___

11. ___C___

# Chapter 9 Answer Key

1. ___D___

2. ___G___

3. ___A___

4. ___H___

5. ___C___

6. ___G___

7. ___C___

8. ___106 ft___

9. ___448 ft²___

10. ___AB = 7 units, BC = 4 units, CD = 7 units, DA = 4 units___

11. ___22 units___

12. ___11 in²___

13. The area is $\frac{1}{2} \cdot \frac{1}{2}$ or $\frac{1}{4}$ times the original area. Area of original figure = 140 ft²; Area of new figure = 35 ft²; 35 ft² ÷ 140 ft² = $\frac{1}{4}$

14. The perimeter is 2 times greater. Perimeter of original figure = 52 ft; Perimeter of new figure = 104 ft; 2 • 52 ft = 104 ft

15. ___64 cm²___

Answers

# Chapter 9 Answer Key

1. ___**B**___

2. ___**H**___

3. ___**D**___

4. ___**F**___

5. ___**B**___

6. ___**H**___

7. ___**C**___

8. ___**140 ft**___

9. ___**781 ft²**___

10. ___*AB* = 3 units, *BC* = 5 units, *CD* = 3 units, *DA* = 5 units___

11. ___**16 units**___

12. ___**15 in²**___

13. The area is $\frac{1}{2} \cdot \frac{1}{2}$ or $\frac{1}{4}$ times the original area. Area of original figure = 32 ft²; Area of new figure = 8 ft²; 8 ft² ÷ 32 ft² = $\frac{1}{4}$

14. The perimeter is 2 times greater. Perimeter of original figure = 28 ft; Perimeter of new figure = 56 ft; 2 · 28 ft = 56 ft

15. ___**110 cm²**___

# Chapter 9 Answer Key

**Test, Form 3A**
**Page 213**

1. **39.2 in²**

2. **12.8 mm**

3. **360 m²**

4. **48 ft²**

5. **46.8 cm**

6. **288 mi²**

7. **The area of the new figure is 352.8 in². This is 9 times the original area, 39.2 in². 352.8 in² ÷ 39.2 in² = 9**

8. **The perimeter is 12 meters. The new perimeter is 48 meters. This is 4 times the original 4 × 12 = 48.**

**Test, Form 3A** *(continued)*
**Page 214**

9. **74 ft²**
Sample answers:
b = 25 units; h = 4 units; b =

10. **20 units; h = 5 units**

11. **76 cm²**

12. **20 ft**

13. **48.5 units²**

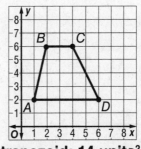

14. **trapezoid; 14 units²**

15. **Sample answer: Subtract the y-coordinates to find the lengths of the vertical sides and subtract the x-coordinates to find the lengths of the horizontal sides.**

**Answers**

# Chapter 9 Answer Key

1. **32 ft²**

2. **11.8 yd**

3. **138.8 in²**

4. **14.3 cm²**

5. **36 ft**

6. **576 cm²**

7. The area of the new figure is 512 ft². This is 16 times the original area, 32 ft². 512 ft² ÷ 32 ft² = 16

8. The original perimeter is 28 feet. The new perimeter is 140 feet. This is 5 times the original. 28 × 5 = 140

9. **166 ft²**

10. Sample answers: $b = 15$ units, $h = 8$ units; $b = 12$ units, $h = 10$ units

11. **144 ft²**

12. **30 ft**

13. **35.5 units²**

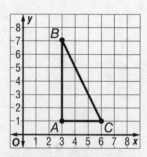

14. **triangle; 9 units²**

15. Sample answer: Subtract the *y*-coordinates to find the lengths of the vertical sides and subtract the *x*-coordinates to find the lengths of the horizontal sides.

# Chapter 10 Answer Key

1. _____ 24

2. _____ 51

3. _____ 4

4. _____ 8

5. _____ 22

6. _____ 30

7. _____ 9

8. _____ 15

9. _____ 40

10. _____ 39

11. _____ 3

12. _____ 55

1. _____ 3

2. _____ 58

3. _____ 10

4. _____ 37

5. _____ 6

6. _____ 6

7. _____ 2

8. _____ 20

9. _____ 25.2

10. _____ 16.2

11. _____ 8.8

12. _____ 10.2

13. _____ 64.8

14. _____ 22.5

15. _____ 202

16. _____ 118.4

17. _____ 20.3

18. _____ 244.8

19. _____ 77

20. _____ 38.4

Answers

# Chapter 10 Answer Key

1. **BASEBALL** John had 30 baseball cards. He gave 14 cards to Mike and 7 to Jeff. Then he bought 6 more cards. How many baseball cards does John have now? **15 cards**

2. **PARTIES** Louise is making a chicken dish for 6 people. The recipe for the chicken dish she is making calls for 1.2 pounds of chicken for each person. Louise wants to double the recipe. How many pounds of chicken will she need? **14.4 pounds**

3. **FUNDRAISER** The soccer team collected soda cans for a fundraiser. They had 175 cans and found 58 more. The next day, they turned in 97 cans. How many cans do they have left? **136**

4. **CORN** Mr. Rodriguez planted 22 rows of corn. There were 15 plants in each row. He also planted 5 rows of tomato plants with each row havin 12 plants. How many plants did he plant in all? **390**

5. **EARNINGS** Max earns $7.25 an hour. If he works for 6 hours a week for 3 weeks, how much will he earn? **$130.50**

6. **WALKING** Val can walk 3.2 miles in an hour. If she walks for 4 hours a week for 5 weeks, how many miles will she walk? **64 mi**

# Chapter 10 Answer Key

**Diagnostic Test**
**Page 220**

1. _____ 43 _____

2. _____ 7 _____

3. _____ 9 _____

4. _____ 21 _____

5. _____ 23 _____

6. _____ 3 _____

7. _____ 19 _____

8. _____ 22 _____

9. _____ 98.4 _____

10. _____ 124.8 _____

11. _____ 231 _____

12. _____ 510.3 _____

13. _____ 38.4 _____

14. _____ 239.4 _____

15. _____ 396 _____

16. _____ $774 _____

17. _____ $660 _____

**Pretest**
**Page 221**

1. _____ 168 m$^3$ _____

2. _____ 285 in$^3$ _____

3. _____ 216 ft$^3$ _____

4. _____ 60 cm$^3$ _____

5. _____ 94 cm$^2$ _____

Answers

# Chapter 10 Answer Key

1. **255 m³**

2. **116.6 in³**

3. **518.4 ft³**

4. **2,200 mm³**

5. **2.76 m³**

6. **7.5 ft**

7. **8 m**

1. **true**

2. **false**

3. **true**

4. **false**

5. **false**

6. **false**

7. **true**

8. **Sample answer: The measure of space a three-dimensional figure occupies.**

9. **Sample answer: Any face that is not a base.**

# Chapter 10 Answer Key

**Student Recording Sheet, Page 226**

*Use this recording sheet with the Standardized Test Practice pages.*

Fill in the correct answer. For gridded-response questions, write your answers in the boxes on the answer grid and fill in the bubbles to match your answers.

1. ● Ⓑ Ⓒ Ⓓ

2. Ⓕ Ⓖ ● Ⓘ

3. Ⓐ Ⓑ ● Ⓓ

4.

5. Ⓕ Ⓖ ● Ⓘ

6. _____ 130.2 cm² _____

7. Ⓐ Ⓑ Ⓒ ●

8.

9. Ⓕ Ⓖ ● Ⓘ

10. _____ **35 in²** _____

11. ● Ⓑ Ⓒ Ⓓ

12. Ⓕ Ⓖ Ⓗ ●

13. Ⓐ ● Ⓒ Ⓓ

14.

15a. 405 in³

15b. It is 8 times greater, 3,240 in³

15c. It makes the new volume twice the original volume. No, it does not matter which dimension is doubled.

## Extended Response

Record your answers for Exercise 15 on the back of this paper.

Answers

# Chapter 10 Answer Key

**Extended-Response Test, Page 227**
**Sample Answers**

*In addition to the scoring rubric, the following sample answers may be used as guidance in evaluating open-ended assessment items.*

**1. a.** $152 \text{ in}^2$

**b.** $96 \text{ in}^3$

**c.** The new volume would be $48 \text{ in}^3$, which is half the original amount. $98 \div 2 = 48$

**2.** 2.5 cm;

Find the area of the base of the prism.

$V = Bh$

$13 = B(4)$

$3.25 = B$

Find the height of the triangle.

$A = \frac{1}{2}bh$

$3.25 = \frac{1}{2}(2.6)h$

$3.25 = 1.3h$

$2.5 = h$

**3. a.**

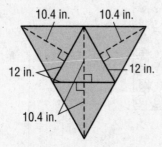

10.4 in.    10.4 in.
12 in.   12 in.
10.4 in.

**b.** To find the surface area of the toy, find the area of one face and then multiply by 4. Since the toy contains equilateral triangles, the base area and lateral areas are equal.

**c.** *S.A.* of each face $= \frac{1}{2}(12)(10.4)$ or 62.4

*S.A.* of toy $= 4(62.4)$ or 249.6

The surface area is 249.6 square inches.

# Chapter 10 Answer Key

1. ___A___

2. ___I___

3. ___B___

4. ___F___

5. ___B___

6. ___G___

7. ___D___

8. ___G___

9. ___C___

10. ___H___

11. ___C___

12. ___H___

**Answers**

# Chapter 10 Answer Key

1. ___B___

2. ___F___

3. ___A___

4. ___G___

5. ___B___

6. ___G___

7. ___D___

8. ___F___

9. ___C___

10. ___F___

11. ___A___

12. ___H___

# Chapter 10 Answer Key

**Test, Form 2A**
**Page 233**

1. _____ C _____

2. _____ G _____

3. _____ B _____

4. _____ G _____

5. _____ B _____

**Test, Form 2A** *(continued)*
**Page 234**

6. _____ F _____

7. _____ D _____

8. _____ 6.5 m _____

9. _____ 166.4 m² _____

10. _____ 308.8 cm² _____

11. _____ 4,320 in³ _____

12. _____ 3.3 cm _____

Answers

# Chapter 10 Answer Key

Test, Form 2B
Page 235

1. _____C_____

2. _____F_____

3. _____B_____

4. _____H_____

5. _____C_____

Test, Form 2B *(continued)*
Page 236

6. _____G_____

7. _____A_____

8. _____9 m_____

9. __480 mm²__

10. __119.0 m²__

11. __10,692 in³__

12. __9.5 cm__

# Chapter 10 Answer Key

**Test, Form 3A**
**Page 237**

1. __266 yd³__

2. __496.8 ft³__

3. __268 m²__

4. __510 ft²__

5. __20 m³__

6. __7 in.__

**Test, Form 3A** *(continued)*
**Page 238**

7. __13 in.__

8. __Box A; 8.25 in³__

9. __6,384 in³; 2,762 in²__

10. __14 m__

11. __5 bags__

12. __yes; the approximate surface area is (30 × 20 × 2) + (30 × 10 × 2) + (20 × 10 × 2) or 2,200 cm².__

**Answers**

# Chapter 10 Answer Key

1. __120 cm³__

2. __582.4 in³__

3. __276 in²__

4. __108 cm²__

5. __88 in²__

6. __6 in.__

7. __16 in.__

8. __Box B;__
   __287.25 in³__

9. __5,832 mm³;__
   __2,268 mm²__

10. __9 m__

11. __5 bags__

12. __yes; the__
   __approximate__
   __surface area is__
   __(40 × 20 × 2) +__
   __(40 × 10 × 2) +__
   __(20 × 10 × 2) or__
   __2,800 cm².__

# Chapter 11 Answer Key

**Are You Ready?—Review**
**Page 241**

1. 44.55

2. 45.79

3. 60.03

4. 82.24

5. 73.51

6. $6.10

7. 62.4 in.

8. $41.23

**Are You Ready?—Practice**
**Page 242**

1. 87.45 mi

2. 74 mi

3. 125.55 mi

4. 15.6

5. 12.5

6. 10.6

7. 10.1

8. 36.3

9. 126.4

10. 141.2 mi

Answers

# Chapter 11 Answer Key

**Are You Ready?—Apply**
**Page 243**

SCHOOL STORE For Exercises 1-3, refer to the table. The table shows the cost of items at the school store.

| School Store | |
|---|---|
| **Item** | **Cost ($)** |
| T-Shirt | 12.95 |
| Sweatshirt | 24.50 |
| Pennant | 8.50 |
| Flag | 14.99 |
| Button | 2.65 |

1. Jessica bought two sweatshirts and two flags. What was her total cost? **$78.98**

2. A new student bought a T-shirt, pennant, flag, and button. How much did the new student spend in all? **$39.09**

3. Suppose you bought one of each of the items in listed in table. What would be the total cost? **$63.59**

4. Three friends decide to split the cost of their bill evenly. If the bill was $65.79, how much will each friend pay? **$21.93**

5. The total weight of five packages of candy is 182.5 ounces. If each package of candy weighs the same amount, how much does one package weigh? **36.5 oz**

6. The total cost of six tickets to a music concert is $208.50. Each ticket costs the same amount. What is the cost of one ticket? **$34.75**

# Chapter 11 Answer Key

**Diagnostic Test**
**Page 244**

1. _____ 79.81 _____

2. _____ 63.81 _____

3. _____ 84.39 _____

4. _____ 89.07 _____

5. _____ 56.19 _____

6. _____ $25.10 _____

7. _____ 91.4 in. _____

8. _____ $19.85 _____

9. _____ $18.85 _____

10. _____ 164.6 mi _____

**Pretest**
**Page 245**

1. _____ 8.4; 10; 11 _____

2. _____ 3.5; 3.5; 4 _____

3. _____ 11; 11; 12 _____

4. _____ 39; 40; 37 _____

5. _____ 55 _____

6. _____ 67.5; 52.5; 85; 32.5 _____

7. _____ 12 _____

8. _____ 11.2 _____

**Answers**

# Chapter 11 Answer Key

## Chapter Quiz
### Page 246

1. _____ $110 _____

2. _ $70.80; $70; $60 _

3. _____ $60 _____

4. _____ 6 years _____

5. ____ 6.4 pounds ____

6. _____ $14.80 _____

7. _____ 66°F _____

8. _____ 64°F _____

9. _____ 60°F _____

## Vocabulary Test
### Page 247

1. _____ median _____

2. _____ mean _____

3. _____ Quartiles _____

4. _____ outliers _____

5. _____ mode _____

6. _____ range _____

7. _____ Average _____

8. measures of center

# Chapter 11 Answer Key

**Student Recording Sheet, Page 250**

*Use this recording sheet with the Standardized Test Practice.*

**Fill in the correct answer. For gridded-response questions, write your answers in the boxes on the answer grid and fill in the bubbles to match your answers.**

1. Ⓐ ● Ⓒ Ⓓ

2. Ⓕ ● Ⓗ Ⓘ

3. Ⓐ Ⓑ ● Ⓓ

4. 
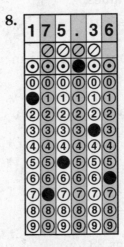

198

5. 

6.85

6. Ⓕ Ⓖ ● Ⓘ

7. Ⓐ ● Ⓒ Ⓓ

8. 

175.36

9. ● Ⓖ Ⓗ Ⓘ

10. _____78_____

11. Ⓐ Ⓑ ● Ⓓ

12. Ⓕ Ⓖ Ⓗ ●

13. ● Ⓑ Ⓒ Ⓓ

14a. 135; 120; none; 195

14b. The inclusion of data value 265 causes the values Part B, to be much higher than if it was not included. The mean would be 102.5, the median would be 102.5, there would be no mode, and the range would be 65.

## Extended Response

Record your answers for Exercise 14 on the back of this paper.

**Answers**

# Chapter 11 Answer Key

### Extended-Response Test, Page 251
### Sample Answers

*In addition to the scoring rubric, the following sample answers may be used as guidance in evaluating extended response assessment items.*

1. **a.** Add the numbers of dollars for all ten weeks, and then divide by 10.
   $$\frac{1077}{10} = 107.70$$

   **b.** List the data from least to greatest. Then find the mean of the middle two values.
   $$\frac{100 + 106}{2} = 103$$

   **c.** No data value appears more than once in the table.

   **d.** Subtract the smallest data value from the greatest data value.
   $$156 - 84 = 72$$

   **e.** upper quartile: 115
   lower quartile: 92
   interquartile
   range: $115 - 92$ or 23

   **f.** Outliers are data values less than the lower quartile minus 1.5 times the interquartile range or greater than the upper quartile plus 1.5 times the interquartile range.

   $92 - 1.5(23) = 57.5$
   $115 + 1.5(23) = 149.5$

   There are no data values less than 57.5. One value, 156, is greater than 149.5. The only outlier is 156.

2. Use the mean to describe a data set when the data have no extreme values. Use the median when the data have extreme values and there are no big gaps in the data. Use the mode when the data have many repeated numbers.

# Chapter 11 Answer Key

**Test, Form 1A**
**Page 253**

1. _____ D _____

2. _____ I _____

3. _____ C _____

4. _____ H _____

5. _____ A _____

6. _____ H _____

7. _____ B _____

**Test, Form 1A** *(continued)*
**Page 254**

8. _____ H _____

9. _____ A _____

10. _____ G _____

11. _____ B _____

12. _____ H _____

13. _____ A _____

**Answers**

# Chapter 11 Answer Key

1. ___A___

2. ___F___

3. ___C___

4. ___I___

5. ___C___

6. ___G___

7. ___C___

8. ___F___

9. ___C___

10. ___H___

11. ___C___

12. ___I___

13. ___B___

# Chapter 11 Answer Key

**Test, Form 2A**
**Page 257**

1. _____ B _____

2. _____ H _____

3. _____ A _____

4. _____ G _____

5. _____ C _____

6. _____ I _____

**Test, Form 2A** *(continued)*
**Page 258**

7. _____ B _____

8. _____ I _____

9. _____ C _____

10. _____ 29.5, 12.5 _____

11. _____ 17 _____

12. _____ 62; 62 is more than 1.5 times the IQR away from the third quartile. _____

13. _____ 5.8 _____

14. _____ 3.3 _____

15. _____ mean; there is no outlier or gap to affect the mean. _____

**Answers**

# Chapter 11 Answer Key

1. _____C_____

2. _____I_____

3. _____A_____

4. _____H_____

5. _____D_____

6. _____G_____

7. _____A_____

8. _____G_____

9. _____D_____

10. _____27, 19_____

11. _____8_____

12. 6; 6 is more than 1.5 times the IQR away from the first quartile.

13. _____5.6_____

14. _____16 in._____

# Chapter 11 Answer Key

Test, Form 3A
Page 261

1. 24.75; 26; no mode

2. median

3. 85, 63

4. 22

5. 0; It is more than 1.5 times the IQR away from the first quartile.

6. 5.5; 3 values

7. 18 in.

8. 18.5 in., 19 in.

Test, Form 3A *(continued)*
Page 262

9. $42

10. with: median; without: mean

11. With the outlier, the mean is 15, the median is 13 and the range is 37. Without the outlier, the mean is 12, the median is 11, and the range is 20. They all decrease. The mode is not affected.

12. $8

13. 3.3

14. 82.5

15. 83, 79

16. mean; There are no outliers or large gaps.

Copyright © The McGraw-Hill Companies, Inc. Permission is granted to reproduce for classroom use.

**Course 1 • Chapter 11** Statistical Measures

# Chapter 11 Answer Key

1. 19.5; 14.5; 12 and 15

2. median

3. 15, 9

4. 6

5. no; There are no values more than 1.5 times the IQR away from the quartiles.

6. 5; 5 values

7. 7.2 lbs

8. 7.3 lbs, 7.4 lbs

9. $18

10. With the outlier, the mean is 5.5, the median is 3, and the range is 17. Without the outlier, the mean is 3, the median is 1 and the rage is 6. They all decrease. The mode is not affected.

11. mode

12. $5.3

13. 1.3

14. 83

15. 83; 78

16. mean; There are no outliers or large gaps.

# Chapter 12 Answer Key

## Are You Ready?—Review
### Page 265

1. _____18_____

2. _____29_____

3. _____33.8_____

4. _____31_____

5. _____8.5_____

## Are You Ready?—Practice
### Page 266

1. _____49_____

2. _____85_____

3. _____8_____

4. _____18_____

5. _____28.8_____

6. _____34.5_____

7. _____128_____

# Chapter 12 Answer Key

1. **GAMES** Jewel likes to play video games. The table shows her scores. What was her average score? Round to the nearest tenth if necessary. **70.2**

   | Jewel's Scores | | |
   | --- | --- | --- |
   | 78 | 69 | 72 |
   | 71 | 71 | 69 |
   | 78 | 72 | 52 |

2. **SPRING** Karly owns a lawn service. During the month of May she averaged 47 customers the first week, 38 the second week, 55 the third week, and 60 the fourth week. How many customers did she average per week? **50 customers**

3. **CANDY** The number of bags of candy sold at a certain movie theater for several days was 32, 17, 20, 38, 47, and 20. Find the average number of bags of popcorn sold during these days. **29 bags**

4. **CUSTOMERS** A store keeps track of the number of customers entering the store in the first hour of each day. In the first four days of June, the store had 215, 143, 100, and 214 customers in the first hour. Find the average number of customers the store had in the first hour during those four days. **168 customers**

5. **KENNELS** Betsy trains dogs. The table shows how many dogs she trained during the first six months this year. What is the median of the data? **19.5**

   | Number of Dogs Trained | | |
   | --- | --- | --- |
   | 21 | 11 | 18 |
   | 23 | 30 | 15 |

6. **CHOIR** Timmy belongs to the municipal choir in his home town. The table below shows the age of each member of the choir. What is the median of the data? **34**

   | Choir Member Ages | | | | | |
   | --- | --- | --- | --- | --- | --- |
   | 18 | 25 | 28 | 33 | 35 | 42 |
   | 24 | 19 | 35 | 47 | 35 | 44 |

# Chapter 12 Answer Key

**Diagnostic Test**
**Page 268**

1. _____ 68 _____

2. _____ 37 _____

3. _____ 30 _____

4. _____ 38 _____

5. _____ 19 _____

6. _____ 23.3 _____

7. _____ 32.6 _____

8. _____ 41.9 _____

9. _____ 62 _____

**Pretest**
**Page 269**

**Roller Coaster Speeds (mph)**

1. _____

2. _____ 25% _____

3. median: 67.5, $Q_1$: 52.5, $Q_3$: 85, range: 55, IQR: 32.5; Sample answer: Both whiskers are approximately the same size so the data is spread evenly below and above the quartiles.

4. _____ 30–39 years _____

5. _____ 20–29 years _____

6. _____ 70–79 years _____

# Chapter 12 Answer Key

## Chapter Quiz
### Page 270

1. **The mode is B.**

2. **4 more students earned a B than an A.**

3. **Student descriptions will vary.**

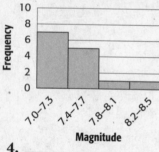

4. _____

5. **2,300**

6. **1,325, 780**

7. **2,450 and 2,800**

**Cost of Computers**

8. _____

## Vocabulary Test
### Page 271

1. **line plot**

2. **gap**

3. **cluster**

4. **distribution**

5. **box plot**

6. **peak**

7. **histogram**

8. **Sample answer: The distribution looks the same on the right and left sides.**

9. **Sample answer: how many data values are in each interval**

# Chapter 12 Answer Key

## Student Recording Sheet, Page 274

*Use this recording sheet with the Standardized Test Practice pages.*

**Fill in the correct answer. For gridded-response questions, write your answers in the boxes on the answer grid and fill in the bubbles to match your answers.**

1. Ⓐ Ⓑ ● Ⓓ

2. Ⓕ ● Ⓗ Ⓘ

3. ● Ⓑ Ⓒ Ⓓ

4. _____ **$16.12** _____

5.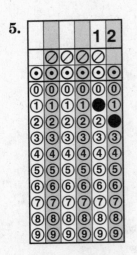

6. Ⓕ Ⓖ Ⓗ ●

7. _____ **25.53 in²** _____

8. Ⓐ Ⓑ ● Ⓓ

9.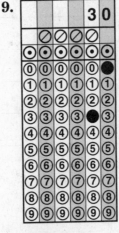

10. Ⓕ ● Ⓗ Ⓘ

11. ● Ⓑ Ⓒ Ⓓ

12. _____ **7 cm** _____

13. Ⓕ ● Ⓗ Ⓘ

14a.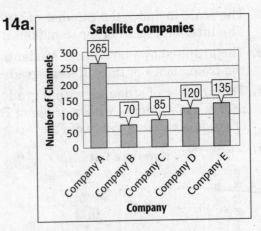

**14b.** 135; 120; none; 195

**14c.** The outlier causes the values to be much higher than if it were not included.
mean = 102.5
median = 102.5
no mode
range = 65

## Extended Response

Record your answers for Exercise 14 on the back of this paper.

**Answers**

# Chapter 12 Answer Key

Extended-Response Test, Page 275
Sample Answers

*In addition to the scoring rubric, the following sample answers may be used as guidance in evaluating extended response assessment items.*

1.

| Students' Grades | | |
|---|---|---|
| Grade | Tally | Frequency |
| A | \|\|\|\| | 4 |
| B | ⊬⊬\| | 6 |
| C | ⊬⊬ ⊬⊬ | 10 |
| D | \|\|\|\| | 4 |
| F | \|\| | 2 |

2. a.

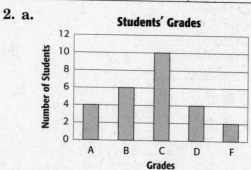

**b.** The scale includes the number of students getting each grade. The interval is 2 in order to make all data fit easily on the graph.

**c.** The graph shows how many students got each grade. It compares the frequencies of the different grades.

**d.** The number of students with a grade of A is the same as the number of students with a grade of D. It is two times the number of students with a grade of F.

3. a.

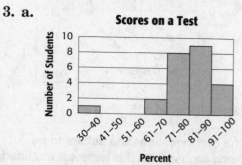

**b.** The data are clustered between 61 and 100, and there is a gap from 41–60.

**c.** mean: 80; median: 82; mode: 82; range: 64

**d.** 34 is an outlier. Without the outlier, the mean is 82. The outlier brings the mean down.

# Chapter 12 Answer Key

**Test, Form 1A**
**Page 277**

1. _____ B _____

2. _____ I _____

3. _____ C _____

4. _____ F _____

5. _____ C _____

6. _____ G _____

7. _____ A _____

**Test, Form 1A** *(continued)*
**Page 278**

8. _____ F _____

9. _____ C _____

10. _____ H _____

11. _____ A _____

12. _____ H _____

**Answers**

# Chapter 12 Answer Key

1. _____**B**_____

2. _____**I**_____

8. _____**H**_____

9. _____**D**_____

3. _____**D**_____

4. _____**G**_____

10. _____**G**_____

11. _____**D**_____

5. _____**B**_____

6. _____**I**_____

12. _____**H**_____

7. _____**B**_____

# Chapter 12 Answer Key

8. _____ I _____

9. _____ C _____

1. _____ B _____

2. _____ G _____

3. _____ B _____

10. _____ F _____

4. _____ G _____

$Q_1$ 8.75, $Q_3$
9.25, MED 8.9,
11. _____ IQR 0.5 _____

5. _____ D _____

12.

**Length of Fish (in.)**

Sample answer:
13. _____ bar graph _____

6. _____ F _____

7. _____ B _____

Answers

# Chapter 12 Answer Key

**Test, Form 2B**
**Page 283**

1.     **A**

2.     **I**

3.     **B**

4.     **F**

5.     **A**

6.     **G**

7.     **D**

**Test, Form 2B** *(continued)*
**Page 284**

8.     **G**

9.     **A**

10.     **F**

11. $Q_1$ 18, $Q_3$ 30, median 23.5, IQR 12

12.

**Number of Toys**

15 20 25 30 35 40

13. Sample answer: bar graph

# Chapter 12 Answer Key

**Test, Form 3A**
**Page 285**

1. _____10_____

2. _____80; 70_____

3. _____75%_____

4. no; The right side is more spread out than the left side.

5. _____median_____

6. interquartile range

7. Sample answer: _____$2,800_____

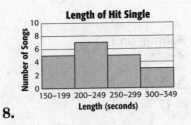

**Length of Hit Single**

8.

**Test, Form 3A**  *(continued)*
**Page 286**

9. _____12_____

10. _____28_____

11. not symmetric; Use the median and interquartile range.

12. $100–$199; mode

13. _____12_____

**Number of Pets**

14.

# Chapter 12 Answer Key

1. _____10_____

2. __80; 70__

3. __75%__

4. no; The left side is more spread out than the right side.

5. __median__

6. interquartile range

7. Sample answer: 1,500 bushels

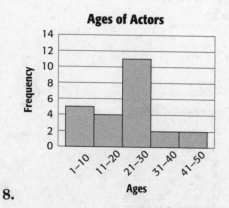

**Ages of Actors**

8.

9. _____30_____

10. _____50_____

11. symmetric; Use the mean and mean absolute deviation.

12. $80–$99; $20–$39

13. _____13_____

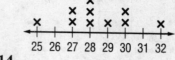

**Annual Rainfall (in.)**

14.

NAME _____ DATE _____ PERIOD _____

## Course 1 Benchmark Test – First Quarter

**1.** The table below shows the number of pennies, nickels, dimes, and quarters that Heather has in her purse. What is the ratio of dimes to nickels expressed as a fraction in simplest form?

| Heather's Coins | |
|---|---|
| Dimes | 9 |
| Nickels | 6 |
| Pennies | 8 |
| Quarters | 4 |

* **A.** $\frac{3}{2}$   **C.** $\frac{3}{5}$

  **B.** $\frac{2}{3}$   **D.** $\frac{2}{5}$

**2.** A cookie recipe calls for a ratio of 4 cups of flour to 3 cups of sugar. For each cup of flour that is used, how many cups of sugar are needed?

**F.** $\frac{4}{3}$ cups of sugar

* **G.** $\frac{3}{4}$ cups of sugar

  **H.** $\frac{2}{3}$ cup of sugar

  **I.** $\frac{3}{7}$ cup of sugar

**3.** Which ratio is *not* equivalent to 5 : 8?

* **A.** 10 out of 18

  **B.** 15 to 24

  **C.** $\frac{20}{32}$

  **D.** 10 : 16

**4. SHORT ANSWER** Mrs. Wilkinson can buy a 20-ounce box of cereal for $3.60 or a 28-ounce box of cereal for $4.20. Which is the better buy? Explain your reasoning.

**28-oz box; The unit rate for the 28-oz box is $0.15 per ounce. The unit rate for the 20-oz box is $0.18 per ounce. 0.15 < 0.20**

**5.** The ratio table shows how much Raymond's brother earns for working different numbers of hours. How many hours would he need to work in order to earn $176?

| Hours | 5 | 8 | 15 | ▓ |
|---|---|---|---|---|
| Earnings | $40 | $64 | $120 | $176 |

**F.** 18 hours

**G.** 20 hours

* **H.** 22 hours

  **I.** 24 hours

**6.** There are 30 students in Mr. Holland's music class. If 30% of the students play in the school band, how many students in the class play in the school band?

* **A.** 9 students

  **B.** 12 students

  **C.** 15 students

  **D.** 100 students

---

NAME _____ DATE _____ PERIOD _____

## Course 1 Benchmark Test – First Quarter *(continued)*

**7. SHORT ANSWER** Jasmine answered 19 out of 25 questions correctly on a quiz. About what percent of her answers were correct? Explain.

**Sample answer: about 80%. $\frac{19}{25}$ is approximately $\frac{20}{25}$, which is 80%.**

**8.** Which of the following shows the rational numbers in order from least to greatest?

**F.** 25%, 0.22, $\frac{1}{5}$

* **G.** $\frac{1}{5}$, 0.22, 25%

  **H.** 0.22, $\frac{1}{5}$, 25%

  **I.** 0.22, 25%, $\frac{1}{5}$

**9.** The model below represents 66%. What decimal is equivalent to this percent?

**A.** 0.066

* **B.** 0.66

  **C.** 6.6

  **D.** 66

**10.** How is the decimal 0.55 written as a fraction in simplest form?

**F.** $\frac{55}{100}$

* **G.** $\frac{11}{20}$

  **H.** $\frac{11}{50}$

  **I.** $\frac{11}{55}$

**11.** Caleb's receipt for lunch is shown below. If Caleb pays with a $10 bill, how much change will he receive?

| Donna's Deli | |
|---|---|
| Chicken Sandwich | $3.79 |
| Soup | $1.45 |
| Drink | $1.29 |
| Tax | $0.39 |

**A.** $6.92

**B.** $5.74

**C.** $3.18

* **D.** $3.08

**12.** Colleen rode her bicycle 9.5 miles in 0.8 hour. What was her average speed in miles per hour?

* **F.** 11.875 miles per hour

  **G.** 10.3 miles per hour

  **H.** 8.7 miles per hour

  **I.** 7.6 miles per hour

**Answers**

NAME _____ DATE _____ PERIOD _____

## Course 1 Benchmark Test – First Quarter (continued)

18. Regina purchased 1.75 pounds of turkey breast from her local deli for $5.99 per pound. To the nearest cent, how much did she spend in all?

   A. $3.42

   B. $7.74

   *C. $10.48

   D. $11.98

19. **SHORT ANSWER** Katrina uses 4.125 yards of fabric for each curtain panel she makes. How many yards will she need if she makes 14 panels?

   **57.75 yards**

20. The table below shows the results of a survey on students' favorite school lunches. What fraction of the students surveyed said that grilled cheese is their favorite school lunch?

   | What Is Your Favorite School Lunch? | |
   | --- | --- |
   | Lunch | Percent |
   | Pizza | 35% |
   | Grilled Cheese | 30% |
   | Spaghetti | 20% |
   | Chicken | 10% |
   | Soup | 5% |

   F. $\frac{3}{100}$

   G. $\frac{1}{5}$

   H. $\frac{1}{4}$

   *I. $\frac{3}{10}$

21. Tommy's batting average this season is 0.275. This means that he had a hit in 27.5% of his at-bats. If Tommy had 11 hits so far this season, how many at-bats has he had?

   *A. 40 at-bats

   B. 35 at-bats

   C. 26 at-bats

   D. 3 at-bats

22. In a machine, a large gear completes a revolution every minute while a small gear completes a revolution every 24 seconds. If the gears are currently aligned, how much time will pass before they are aligned again?

   F. 12 seconds

   G. 24 seconds

   H. 1 minute

   *I. 2 minutes

23. Which of the following is the best estimate for the problem shown below?

   $151 \div 29$

   A. about 4

   *B. about 5

   C. about 6

   D. about 7

292

---

NAME _____ DATE _____ PERIOD _____

## Course 1 Benchmark Test – First Quarter (continued)

13. There are 25 servings in a 30.2-ounce jar of peanut butter. How many ounces of peanut butter are there in 1 serving?

   *A. 1.208 ounces

   B. 1.15 ounces

   C. 0.828 ounce

   D. 0.64 ounce

14. **SHORT ANSWER** Angela earns $6.25 per hour babysitting. Estimate how much she will earn if she babysits for 9 hours this weekend. Explain.

   **Sample answer: about $54. Round $6.25 to $6 and multiply by 9.**

15. Which of the following is *not* a factor of 84?

   F. 2

   G. 3

   H. 7

   *I. 8

16. Which of the following does *not* represent the shaded portion of the figure below?

   *A. 25%

   B. $\frac{3}{4}$

   C. 75%

   D. 0.75

17. Roberta uses 5 beads for every 3 inches of string while making necklaces. Which graph best represents the ratio of necklace string to beads used?

   F.

   G.

   H.

   *I.

291

NAME _____ DATE _____ PERIOD _____

## Course 1 Benchmark Test – Second Quarter

1. Raul is making a scale model of an airplane that has a wingspan of 44 feet. If Raul's scale model is $\frac{1}{16}$ the size of the actual airplane, what is the wingspan of his model?

   A. 704 ft

   B. 60 ft

   *C. $2\frac{3}{4}$ ft

   D. $1\frac{2}{3}$ ft

2. Two-thirds of the students in Hannah's homeroom plan to do some volunteering this summer. Of these students, $\frac{3}{5}$ plan to volunteer at the community center. What fraction of the students in Hannah's homeroom plan to volunteer at the community center this summer?

   F. $\frac{2}{3}$

   G. $\frac{3}{5}$

   *H. $\frac{2}{5}$

   I. $\frac{1}{15}$

3. **SHORT ANSWER** Which point on the number line is closest to the product of the numbers graphed at points $R$ and $T$? Explain your answer.

   point Q; Point $R = \frac{7}{10}$ and Point
   $T = \frac{9}{10}, \frac{7}{10} \cdot \frac{9}{10} = \frac{63}{100}$ and $\frac{63}{100} \approx$
   $\frac{6}{10}$, point Q.

4. In which quadrant does point A lie on the coordinate plane?

   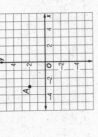

   A. I

   *B. II

   C. III

   D. IV

5. Which of the following integers has the greatest absolute value?

   F. 0

   G. 7

   *H. −10

   I. 1

6. The Panthers football team lost 4 yards on each of their first two plays of the game. Which of the following integers represents the progress of the team after the first two plays?

   *A. −8

   B. −4

   C. 4

   D. 8

---

NAME _____ DATE _____ PERIOD _____

## Course 1 Benchmark Test – First Quarter (continued)

24. **SHORT ANSWER** One acre is equivalent to 43,560 square feet. If an acre is also equivalent to 4,840 square yards, how many square feet are equal to one square yard?

    **9 square feet**

25. Which model represents 140%?

    F.

    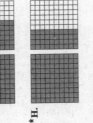
    G.

    *H.

    I.

**Answers**

NAME _____ DATE _____ PERIOD _____

## Course 1 Benchmark Test – Second Quarter *(continued)*

**7.** The table shows the record low temperatures of four different towns. Which of the following shows the record temperatures ordered from least to greatest?

| Record Low Temperatures | |
|---|---|
| Town | Temperature (°F) |
| Oakmont | −7 |
| Cherry Grove | 3 |
| Anderson Hills | 11 |
| Glentown | −2 |

**F.** 11, 3, −2, −7

**G.** −2, 3, −7, 11

**H.** −2, −7, 3, 11

**\*I.** −7, −2, 3, 11

**8.** Which of the following expressions correctly uses exponents to show the prime factorization of 360?

**A.** $2^4 \times 3^2 \times 5$

**\*B.** $2^3 \times 3^2 \times 5$

**C.** $2^4 \times 3 \times 5$

**D.** $2^3 \times 3 \times 5^2$

**9.** The expression $\frac{d}{t}$ can be used to find the average speed of an object that travels a distance $d$ in time $t$. What is a car's average speed if it travels 145 miles in 2.5 hours?

**\*F.** 58 miles per hour

**G.** 62 miles per hour

**H.** 65 miles per hour

**I.** 362.5 miles per hour

**10.** Which of the following expressions is equivalent to $6(5 + 3x)$?

**A** $30 + 3x$

**B** $11 + 9x$

**\*C** $30 + 18x$

**D** $11 + 3x$

**11. SHORT ANSWER** Graph and label point $W(4, -1)$ on the coordinate plane.

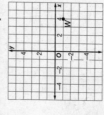

**12.** What are the coordinates of the point in Quadrant IV on the coordinate plane?

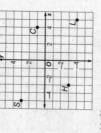

**F.** (4, 1)

**G.** (1, 4)

**H.** (−4, 5)

**\*I.** (5, −4)

NAME _____ DATE _____ PERIOD _____

## Course 1 Benchmark Test – Second Quarter *(continued)*

**13.** Which of the following rational numbers represents a repeating decimal?

**\*A.** $\frac{25}{48}$

**B.** $\frac{11}{40}$

**C.** $\frac{7}{32}$

**D.** $\frac{3}{25}$

**14.** The top students in a distance throwing competition are shown in the table. How many yards did the winner of the competition throw the ball?

| Distance Throwing Competition | |
|---|---|
| Student | Distance (ft) |
| Ashley | 162 |
| Craig | 156 |
| Fernando | 175 |
| Robert | 166 |

**F.** 525 yards

**G.** 468 yards

**\*H.** $58\frac{1}{3}$ yards

**I.** 52 yards

**15. SHORT ANSWER** Define a variable and write an expression to represent the following phrase.

*seven years younger than Lisa*

**Let a represent Lisa's age; a − 7**

**16.** Mrs. Rome has $\frac{2}{3}$ of a pan of lasagna left after dinner. She wants to divide the leftover lasagna into 4 equal servings. What fraction of the original pan does each serving represent?

**A.** $\frac{1}{12}$

**\*B.** $\frac{1}{6}$

**C.** $\frac{1}{4}$

**D.** $\frac{3}{8}$

**17.** Jeff is making fruit punch for the school dance. He needs $3\frac{3}{4}$ cups of pineapple juice per batch. If Jeff wants to make $4\frac{1}{2}$ batches of punch, how many cups of pineapple juice will he need?

**F.** $8\frac{1}{4}$ cups

**G.** $12\frac{3}{8}$ cups

**H.** $15\frac{1}{2}$ cups

**\*I.** $16\frac{7}{8}$ cups

**18.** Which of the following symbols, when placed in the blank, makes the number sentence true?

$$\frac{11}{12} \underline{\hspace{1cm}} 0.916666...$$

**A.** +

**\*B.** =

**C.** <

**D.** >

# Benchmark Test Answer Keys

NAME _____ DATE _____ PERIOD _____

## Course 1 Benchmark Test – Second Quarter *(continued)*

**19. SHORT ANSWER** A kindergarten teacher has $22\frac{1}{2}$ cups of juice to be divided equally among her students. If each student is to receive $1\frac{1}{4}$ cups of juice, how many students are there?

**18 students**

**20.** A plumber has 28 feet of PVC pipe that he wants to cut into sections that are $2\frac{1}{3}$ feet long. How many sections of pipe will the plumber have in all?

F. $14\frac{1}{3}$ sections

G. $13\frac{1}{2}$ sections

*H. 12 sections

I. 11 sections

**21.** Which property is represented by the equation below?

$$\frac{2}{3} \times \frac{3}{2} = 1$$

*A. Multiplicative Inverse Property

B. Multiplicative Identity Property

C. Distributive Property

D. Commutative Property of Multiplication

**22.** Alexandria is evaluating the expression below.

$$3 \times 8 \div 2 + (4 - 1)^2$$

Which operation should be performed first according to the order of operations?

F. Multiply 3 and 8.

G. Divide 8 by 2.

*H. Subtract 1 from 4.

I. Evaluate the power.

**23.** Which of the following coordinate planes correctly shows point $G(4, -5)$ graphed?

A.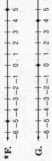

B.

C.

*D.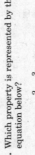

NAME _____ DATE _____ PERIOD _____

## Course 1 Benchmark Test – Second Quarter *(continued)*

**24.** Which number line shows two different integers with the same absolute value?

*F.

G.

H.

I.

**25. SHORT ANSWER** Use the Distributive Property to write a numerical expression that is equivalent to $25 + 10$.

**5(5 + 2)**

**Answers**

# Benchmark Test Answer Keys

NAME _____ DATE _____ PERIOD _____

## Course 1 Benchmark Test – Third Quarter

1. The algebra mat below models the equation $x + 3 = 5$.

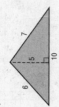

What is the solution to the equation?

A. $-8$

B. $-2$

*C. 2

D. 8

2. The sixth graders at Lakota Middle School hope to raise *at least* $250 for a local charity. Which of the following inequalities best represents this situation?

*F. $n \geq 250$

G. $n > 250$

H. $n < 250$

I. $n \leq 250$

3. **SHORT ANSWER** Paco's age divided by 3 is equal to 4. Let $p$ represent Paco's age and write an equation to model the situation. Then solve the equation to find Paco's age.

$\dfrac{p}{3} = 4$; 12 years old

4. What is the area of the triangle shown below?

A. 23 square units

*B. 25 square units

C. 42 square units

D. 50 square units

5. Which number line shows the solution to the inequality $x - 1 > 2$?

F.

G.

H.

*I.

6. Which operation should be performed to solve the equation below?

$$x - 8 = 11$$

A. Divide each side by 8.

B. Multiply each side by 8.

*C. Add 8 to each side.

D. Subtract 8 from each side.

**Course 1** • Benchmark Test – Third Quarter          **299**

---

NAME _____ DATE _____ PERIOD _____

## Course 1 Benchmark Test – Third Quarter (continued)

7. Grant earns 10 points for each bull's-eye that he hits in an archery competition. Which equation shows the relationship between the number of bull's-eyes Grant hits $b$ and the number of points earned $n$?

F. $n = \dfrac{b}{10}$

*G. $n = 10b$

H. $b = 10n$

I. $b = \dfrac{10}{n}$

8. Joshua has a rectangular vegetable garden with the dimensions shown below.

If he doubles the length and width of the garden, how will the area of the new garden compare to the area of the original garden?

*A. The area of the new garden will be 4 times larger.

B. The area of the new garden will be 2 times larger.

C. The area of the new garden will be half as large.

D. The area of the new garden will be one fourth as large.

9. What value of $x$ results in a true number sentence in the equation?

$$\frac{x}{6} = 2$$

F. 3

G. 4

H. 8

*I. 12

10. What is the missing rule in the function table?

| x | ? |
|---|---|
| 0 | 0 |
| 1 | 1 |
| 3 | 9 |
| 4 | 16 |
| 5 | 25 |

A. $-3x$

B. $-2x$

C. $2x$

*D. $x^2$

11. **SHORT ANSWER** Complete the function table shown below.

| Input (x) | Output (x + 8) |
|-----------|----------------|
| 1 | 9 |
| 4 | 12 |
| 7 | 15 |
| 10 | 18 |

12. Which inequality is graphed on the number line?

*F. $x > -3$

G. $x \geq -3$

H. $x \leq -3$

I. $x < -3$

**Course 1** • Benchmark Test – Third Quarter          **300**

NAME _____ DATE _____ PERIOD _____

## Course 1 Benchmark Test – Third Quarter  (continued)

**13.** SHORT ANSWER What is the area of the shaded parallelogram shown below? Explain your answer.

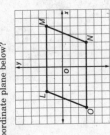

**16 square units; The formula for the area of parallelogram is $A = bh$. Since $b = 4$ units and $h = 4$ units, then the area is equal to $4 \times 4$ or 16.**

**14.** Cameron earns $6 per hour walking dogs in his neighborhood. Which of the following does *not* represent the relationship between the number of hours $x$ and Cameron's earnings $y$?

A. Input: $x$; Output: $6x$

*B. $x = 6y$

C.

| x | 1 | 2 | 3 | 4 |
|---|---|---|---|---|
| y | 6 | 12 | 18 | 24 |

D.

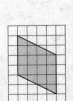

**15.** Eric spent $24 on raffle tickets. If each ticket cost $3, which equation could be used to find the number of tickets $t$ that Eric purchased?

*F. $3t = 24$

G. $24t = 3$

H. $\dfrac{t}{3} = 24$

I. $\dfrac{t}{24} = 3$

**16.** What type of polygon is shown on the coordinate plane below?

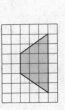

A. rectangle

B. square

C. rhombus

*D. parallelogram

**17.** What is the area of the trapezoid shown below?

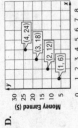

F. 24 square units

G. 15 square units

*H. 12 square units

I. 9.5 square units

NAME _____ DATE _____ PERIOD _____

## Course 1 Benchmark Test – Third Quarter  (continued)

**18.** What is the area of the composite figure shown below?

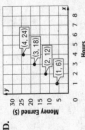

A. 31 square units

B. 28.5 square units

*C. 26 square units

D. 22.5 square units

**19.** Which property of equality would you use to solve the equation?

$$\frac{w}{5} = -10$$

F. Addition Property of Equality

G. Division Property of Equality

H. Subtraction Property of Equality

*I. Multiplication Property of Equality

**20.** Which rule best describes the relationship shown in the function table?

| Input | Output |
|-------|--------|
| 4 | 7 |
| 8 | 11 |
| 12 | 15 |
| 16 | 19 |
| 20 | 23 |

A. subtract 3

*B. add 3

C. divide by 2

D. multiply by 2

**21.** Crystal's dog and puppy weigh 102 pounds combined. If Crystal's dog weighs 87 pounds, which equation could be used to find the weight in pounds the puppy $p$?

F. $\dfrac{p}{87} = 102$

G. $87p = 102$

H. $87 - p = 102$

*I. $87 + p = 102$

**22.** SHORT ANSWER Meredith burns 12 calories per minute while jogging. At this rate, how long will it take her to burn 180 calories? Write and solve an equation to represent this situation.

**$12m = 180$; 15 minutes**

**23.** Which equation represents the function graphed on the coordinate plane?

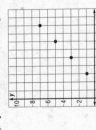

A. $y = x + 2$

*B. $y = x - 2$

C. $x = y - 2$

D. $x = 2 - y$

**Answers**

## Course 1 Benchmark Test – Third Quarter (continued)

NAME _____ DATE _____ PERIOD _____

**24.** Coach Wilson hopes to have *no more than* 15 turnovers during today's basketball game. Which inequality best represents this situation?

**F.** $t \geq 15$

**G.** $t > 15$

**H.** $t < 15$

***I.** $t \leq 15$

**25.** SHORT ANSWER Draw and label a triangle that has an area of 40 square units.

**Sample answer:**

---

## Course 1 Benchmark Test – End of Year

NAME _____ DATE _____ PERIOD _____

**1.** Which rule best describes the relationship shown in the function table below?

| Input | Output |
|-------|--------|
| 1 | 3 |
| 2 | 6 |
| 3 | 9 |
| 4 | 12 |
| 5 | 15 |

**A.** subtract 2

**B.** add 2

**C.** divide by 3

***D.** multiply by 3

**2.** Marcus needs to earn a grade *higher than* 88 on his final quiz in order to have an A average. Which inequality best represents this situation?

**F.** $g \geq 88$

***G.** $g > 88$

**H.** $g < 88$

**I.** $g \leq 88$

**3.** SHORT ANSWER Define a variable and write an expression to represent the following phrase.

*a number increased by 5*

**Let *n* represent the number; $n + 5$**

**4.** What is the least common multiple of 8 and 14?

***A.** 56

**B.** 28

**C.** 4

**D.** 2

**5.** What is the volume of the rectangular prism shown below?

5 cm
3 cm
12 cm

**F.** 20 cm³

**G.** 75 cm³

***H.** 180 cm³

**I.** 222 cm³

**6.** The list below shows the number of books read by students in Abram's class over the summer. What is the mode of the data?

3, 6, 12, 4, 3, 5, 4, 8, 4, 10, 4, 8, 7, 5, 7

***A.** 4 books

**B.** 5 books

**C.** 7 books

**D.** 9 books

NAME _____ DATE _____ PERIOD _____

## Course 1 Benchmark Test – End of Year (continued)

7. Which type of data display would be best for showing how data change over time?

F. box plot

G. histogram

*H. line graph

I. line plot

8. There are 65 people watching a movie at a theater. If 40% of the customers purchased refreshments for the movie, how many customers purchased refreshments?

*A. 26 customers

B. 34 customers

C. 39 customers

D. 163 customers

9. Adeline is wrapping a gift for her mother in a box with the dimensions shown.

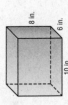

What is the minimum amount of wrapping paper Adeline will need to completely cover the gift box?

F. 188 square inches

*G. 376 square inches

H. 424 square inches

I. 488 square inches

10. The ratio table shows the number of miles Karen can drive for 1, 2, 3, and 4 gallons of gasoline. Based on the table, how far would she be able to drive on 8 gallons of gasoline?

| Gallons | 1 | 2 | 3 | 4 |
|---|---|---|---|---|
| Distance (mi) | 30 | 60 | 90 | 120 |

A. 30 mi

B. 150 mi

C. 210 mi

*D. 240 mi

11. SHORT ANSWER Emily made 14 out of 19 shots during basketball practice. About what percent of her shots did she make? Explain your reasoning.

Sample answer: about 70%. $\frac{14}{19}$ is approximately $\frac{14}{20}$, which is 70%.

12. A muffin recipe calls for a ratio of 5 cups of flour to 2 cups of sugar. For each cup of sugar that is used, how many cups of flour are needed?

*F. $\frac{5}{2}$ cups of flour

G. $\frac{5}{7}$ cups of flour

H. $\frac{2}{5}$ cup of flour

I. $\frac{2}{7}$ cup of flour

Course 1 • Benchmark Test – End of Year

---

NAME _____ DATE _____ PERIOD _____

## Course 1 Benchmark Test – End of Year (continued)

13. SHORT ANSWER The line graph shows the number of members during the first few months of a photography club. Describe the data. Then predict the number of members for the sixth month.

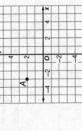

The number of members is steadily increasing. If the trend continues, there will be about 13 or 14 members in the sixth month.

14. The table shows the number of points Anna scored this season. Find the mean number of points Anna scored.

| Points Scored | | | |
|---|---|---|---|
| 12 | 7 | 9 | 10 |
| 16 | 6 | 8 | 15 |
| 12 | 11 | 12 | 14 |

A. 9 points

B. 10 points

*C. 11 points

D. 12 points

15. Which of the following integers has the least absolute value?

*F. –3

G. 4

H. 8

I. –12

16. Albert purchased 2.4 pounds of mixed nuts for $4.79 per pound. How much did he spend in all, to the nearest cent?

A. $12.43

*B. $11.50

C. $6.71

D. $1.99

17. Which of the following coordinate pairs corresponds to point A?

F. (2, –3)

G. (3, –2)

H. (–2, 3)

*I. (–3, 2)

18. Which of the following symbols, when placed in the blank, makes the number sentence true?

$$\frac{20}{75} \ \_\_\_ \ 0.26$$

A. +

B. =

C. <

*D. >

Course 1 • Benchmark Test – End of Year

Answers

NAME _____ DATE _____ PERIOD _____

## Course 1 Benchmark Test – End of Year (continued)

**19.** What is the volume of the triangular prism?

8 m  6 m  14 m  6 m

*F. 336 cubic meters

G. 384 cubic meters

H. 672 cubic meters

I. 724 cubic meters

**20.** The line plot shows the quiz scores of several students.

Quiz Scores

0 1 2 3 4 5 6 7 8 9 10

What is the range of the quiz scores?

A. 4 points

*B. 5 points

C. 7 points

D. 8 points

**21.** Julio is evaluating the expression below.

$$6 + 2(9 - 4) - 3 \times 5$$

Which operation should be performed first according to the order of operations?

F. Add 6 and 2.

G. Multiply 2 by 9.

*H. Subtract 4 from 9.

I. Multiply 3 by 5.

**22.** Which property is represented by the equation shown below?

$$6 \times 3 = 3 \times 6$$

A. Multiplicative Inverse Property

B. Multiplicative Identity Property

C. Associative Property of Multiplication

*D. Commutative Property of Multiplication

**23.** The algebra mat below models the equation $x - 2 = 4$.

What is the solution to the equation?

*F. 6

G. 2

H. −2

I. −8

**24.** SHORT ANSWER Graph the figure with the vertices $A(2, -1)$, $B(6, -1)$, and $C(6, 4)$. Then classify the figure.

**right triangle**

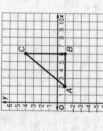

---

NAME _____ DATE _____ PERIOD _____

## Course 1 Benchmark Test – End of Year (continued)

**25.** Which number line shows the solution to the inequality $x + 3 \le 1$?

A. −5 −4 −3 −2 −1 0 1 2 3 4 5

B. −5 −4 −3 −2 −1 0 1 2 3 4 5

*C. −5 −4 −3 −2 −1 0 1 2 3 4 5

D. −5 −4 −3 −2 −1 0 1 2 3 4 5

**26.** The box plot shows the daily attendance at a fitness class.

Fitness Class Attendance

20 25 30 35 40 45 50 55 60

What is the median of the data?

F. 55

*G. 40

H. 35

I. 20

**27.** What value of $x$ results in a true number sentence in the equation shown?

$$2x = 16$$

A. 32

B. 14

*C. 8

D. 4

**28.** Which of the following equations represents the function graphed on the coordinate plane?

F. $y = x + 5$

G. $y = x + 1$

H. $y = 11 + x$

*I. $y = 11 - x$

**29.** SHORT ANSWER The table below shows computer prices at an electronics store.

| Computer Prices ($) | | | |
|---|---|---|---|
| 950 | 620 | 545 | 810 |
| 775 | 1,120 | 905 | 775 |

Find the mean absolute deviation to the nearest cent. Explain what this value represents.

**$134.38; This means that the average distance each data value is from the mean is $134.38.**

NAME _____ DATE _____ PERIOD _____

## Course 1 Benchmark Test – End of Year *(continued)*

**36.** Kylie surveyed several classmates about the number of states they have visited. The results are shown in the histogram.

**How Many States Have You Visited?**

(histogram: Frequency vs Number of States, categories 0–3, 4–7, 8–11, 12–15, 16–19, 20–23, 24–27)

How many of Kylie's classmates have visited more than 15 states?

**F.** 3 students

**G.** 8 students

***H.** 12 students

**I.** 15 students

**37.** What is the surface area of the triangular prism?

6 cm, 7 cm, 9 cm, 8 cm, 15 cm

**A.** 468 square centimeters

***B.** 414 square centimeters

**C.** 405 square centimeters

**D.** 378 square centimeters

**38.** Which of the following represents the decimal 0.32 written as a fraction in simplest form?

**F.** $\frac{32}{100}$

**G.** $\frac{16}{50}$

**H.** $\frac{17}{50}$

***I.** $\frac{8}{25}$

**39.** **SHORT ANSWER** Jeremy can purchase a 1.2-pound package of ground beef for $4.55 or a 1.6-pound package for $6.30. Which is the better buy? Explain your reasoning.

**1.2-pound; The 1.2-pound package has a unit rate of about $3.79 per pound. The 1.6-pound package has a unit rate of about $3.94 per pound. 3.79 < 3.94**

**40.** Pamela is the leading server on her volleyball team. On average, she serves an ace 44% of the time. If she attempts 25 serves in her next game, how many aces would you expect her to have?

**A.** 57 aces

**B.** 19 aces

***C.** 11 aces

**D.** 8 aces

Course 1 • Benchmark Test – End of Year

310

---

NAME _____ DATE _____ PERIOD _____

## Course 1 Benchmark Test – End of Year *(continued)*

**30.** The table below shows the type and number of vehicles in a parking lot.

| Types of Cars | |
| --- | --- |
| Minivans | 12 |
| Sedan | 28 |
| SUV | 9 |
| Trucks | 5 |

What is the ratio of sedans to minivans in simplest form?

***A.** 7 to 3

**B.** 3 to 7

**C.** 7 to 10

**D.** 10 to 3

**31.** The expression $rt$ can be used to find the distance traveled by an object that has an average speed of $r$ over time $t$. How many miles will a hot air balloon travel in 2.2 hours if it travels at an average speed of 12.5 miles per hour?

**F.** 30.1 miles

***G.** 27.5 miles

**H.** 14.7 miles

**I.** 5.7 miles

**32.** What is the area of the triangle?

(right triangle with sides 15, 8, 17)

**A.** 120 square units

**B.** 75 square units

***C.** 60 square units

**D.** 40 square units

**33.** **SHORT ANSWER** The table below shows the number of canoes rented from Outdoor Adventures over the past four weekends.

| Canoe Rentals | | | |
| --- | --- | --- | --- |
| 21 | 32 | 17 | 24 |
| 15 | 30 | 28 | 26 |

Find the range, median, first quartile, third quartile, and interquartile range of the data.

**range: 17, median: 25, first quartile: 19, third quartile: 29, interquartile range: 10**

**34.** A carpenter makes 4 table legs for each table that he builds. Which equation represents the relationship between the number of tables built $t$ and the number of legs made $l$?

***F.** $l = 4t$

**G.** $t = 4l$

**H.** $l = t + 4$

**I.** $t = l + 4$

**35.** Which of the following ratios is equivalent to $\frac{5}{8}$?

**A.** 16 : 10

**B.** 5 to 13

**C.** $\frac{25}{44}$

***D.** 15 out of 24

Course 1 • Benchmark Test – End of Year

309

---

**Answers**

NAME _____ DATE _____ PERIOD _____

## Course 1 Benchmark Test – End of Year  *(continued)*

**41.** What percent is represented by the model?

*F. 175%

G. 125%

H. 75%

I. 25%

**42.** Which of the following best describes the center of a data set if there are outliers in the data but no big gaps in the middle of the data?

A. mean

*B. median

C. mode

D. range

**43. SHORT ANSWER** Complete the function table.

| Input (x) | Output (3x − 1) |
|---|---|
| 1 | 2 |
| 2 | 5 |
| 3 | 8 |
| 4 | 11 |
| 5 | 14 |

**44.** What is the surface area of a square pyramid with base side lengths of 10 inches and a slant height of 14 inches?

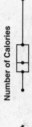

F. 220 in²

G. 280 in²

*H. 380 in²

I. 660 in²

**45.** Which of the following properties would you use to solve the equation?

$$r + 4 = 11$$

A. Addition Property of Equality

B. Division Property of Equality

C. Multiplication Property of Equality

*D. Subtraction Property of Equality

**46.** Which of the following inequalities is graphed on the number line?

F. x > 4

G. x ≥ 4

*H. x ≤ 4

I. x < 4

NAME _____ DATE _____ PERIOD _____

## Course 1 Benchmark Test – End of Year  *(continued)*

**47. SHORT ANSWER** The box plot below shows the number of Calories in different lunches at a restaurant. Describe the shape of the distribution using symmetry and outliers.

**Number of Calories**

**There is an outlier at 275 Calories. The right whisker is longer than the left whisker, and the right box is longer than the left box. So, the data is not symmetric. It is skewed right.**

**48.** What is the area of trapezoid *QRST*?

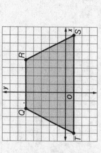

*A. 54 square units

B. 68 square units

C. 76 square units

D. 108 square units

**49.** Mr. Addison is building a sandbox shaped like a rectangular prism. The sandbox is 8 feet long, 6 feet wide, and 1.5 feet deep. How many cubic feet of sand will the sandbox hold?

F. 15.5 cubic feet

*G. 72 cubic feet

H. 105 cubic feet

I. 138 cubic feet

**50.** The Pirates football team has played 75% of its games so far this season. If the team has played 9 games, how many games are there in the season?

A. 7 games

B. 11 games

*C. 12 games

D. 15 games

**51.** Which of the following expressions is equivalent to 3(4x + 1)?

F. 7x + 4

G. x + 4

H. 12x + 1

*I. 12x + 3

**52.** What is the missing rule in the function table?

| x | ? |
|---|---|
| 2 | 7 |
| 3 | 8 |
| 6 | 11 |
| 9 | 14 |
| 12 | 17 |

A. $\dfrac{x}{-4}$

*B. x + 5

C. −4x

D. x − 3

NAME _____ DATE _____ PERIOD _____

# Course 1 Benchmark Test – End of Year *(continued)*

**53.** Which of the following expressions correctly uses exponents to show the prime factorization of 168?

   **F.** $2^4 \times 3 \times 7$

   **G.** $2^3 \times 3^2 \times 7$

   **H.** $2^4 \times 3^2 \times 7$

   *** I.** $2^3 \times 3 \times 7$

**55.** A pancake recipe calls for $\frac{1}{3}$ cup of mix for 4 pancakes. If Beth needs to make 60 pancakes, how many cups of pancake mix will she need?

   ***A.** 5 cups

   **B.** $4\frac{2}{3}$ cups

   **C.** $3\frac{1}{3}$ cups

   **D.** $\frac{1}{5}$ cup

**54. SHORT ANSWER** Which measure of center would you use to describe the center of the data shown on the line plot? Explain your reasoning.

**Number of Pets**

```
                      x x
        x           x x
        x x x x x   x x
        x x x x x   x x
      x x x   x x   x x
      x x x   x x   x x
      1 2 3 4 5 6 7 8 9 10
```

**mean; There are no extreme values in the data set.**

313

**Answers**

**A157**